Water Pollution and Toxicology

Editors
S.K. Shukla
P.R. Srivastava

Commonwealth

Water Pollution and Toxicology

ISBN : 81-7169-204-4

Edition : 2026

Published by :
COMMONWEALTH PUBLISHERS (P) LTD.
4831/24, Govind Lane,
Ansari Road, Darya Ganj,
New Delhi - 110002
Ph. : 23272541, 23242541
e-mail : commonwealthpublishers2016@gmail.com

Distributed by :
ARJUN PUBLISHING HOUSE
4831/24, Govind Lane,
Ansari Road, Darya Ganj,
New Delhi - 110002
Ph. : 23272541, 23242541
e-mail : arjunbooks2001@gmail.com

Laser Typesetting by :
Dimple Computers
Delhi

Printed at :
Roshan Digital Printers
Delhi

PREFACE

Toxicology is the quantitative study of the harmful effects of chemicals on organisms. The field of toxicology is a broad one and may be broken down into three major divisions, namely economic forensic, and environmental toxicology. Economic toxicology is concerned with the deliberate use of toxic chemicals to produce harmful effects on target organisms such as bacteria parasite and insects. Environmental toxicology is concerned with the incidental exposure of plants and animals including humans, to chemicals that are contaminants of environment including in particular the water we drink and the water in which aquatic organisms live. Environmental toxicology is therefore directly related to the subject of water pollution because it is environmental toxicological studies that reveals the quantitative relationships between the concentration of organisms and the effects of these chemicals on aquatic organisms and on persons who drink the water or consumes the organisms.

EDITORS

CONTENTS

1

INTRODUCTION

Outside of a laboratory pure water does not exist, and it would be uneconomic to attempt to produce it. The quality of water is therefore judged in relation to the purpose for which it is to be used, as shown in Table 1.1.

The relationship between water quality and human activities is extremely complicated. Water is used extensively for domestic, industrial, and agricultural purposes and after use is usually returned in a degraded state to rivers, lakes, estuaries, or oceans. Wastes also find their way by leaching into groundwater or by direct discharge into the estuaries or the sea. Industrial and agricultural operations such as smelting or crop spraying introduce materials into the aquatic environment either directly or via the atmosphere. All stages of the hydrological cycle are therefore subject to human activity. Not all of the activity causes a deleterious change in quality. In treatment of water and wastewaters the quality is deliberately improved, and improvements may result incidentally from other activities such as the construction of reservoirs.

The control of water pollution is therefore essential if water resources are to be properly managed and damage to wildlife, recreation, crops, and human populations is to be avoided. Some legislative and administrative framework is required, which implies the setting of standards or guidelines for quality and some framework of monitoring to ensure that the standards are being achieved.

TABLE 1.1

Relationship Between Water Usage and Water Quality

Water use	*Quality requirement*
Navigation	Free from large masses of floating debris (*e.g.*, vegetation), which may foul propellors
Power generation	As above, to avoid damage to intake structure, plus inert suspended solids limitation to prevent erosion of turbines
Recreation	Limit on organic content to prevent anaerobic conditions arising: turbidity to give light penetration for viewing; suspended solids to avoid unsightly deposits; oil, grease, or other floating matter, which renders the surface unsightly; nutrients to prevent undesirable growth developing; fecal contamination to prevent risk of spreading intestinal and cutaneous disease
Fishing	Limit on organic content to prevent deoxygenation; turbidity to permit plant growth and photosynthesis; suspended solids to avoid damage to the benthic community especially near breeding grounds; oil, grease, and other floating matter for affecting recreaction capacity; toxic matter, which may affect members of the aquatic community in the short or long term; nutrients to prevent undesirable growths that may cause marked diurnal fluctuations in DO or by altering the nature of the substrate change the flora and fauna
Abstractions for domestic supply	Limit on faecal contamination to avoid spreading intestinal diseases; toxic materials to prevent chemical poisoning; substances affecting palatability such as taste, odour, salinity, colour, turbidity, solids; hardness

There are difficulties in producing surveillance programmes. These are partly conceptual since no completely satisfactory parameters can be found to measure the esthetic qualities of water such as taste and odour. There are also difficulties in choosing sampling sites, sampling frequency, and so forth. Nevertheless, monitoring is an essential part of any water pollution control programme.

In discussing water quality it is important to distinguish between abstracted waters and natural waters. The quality of abstracted waters, particularly where they are intended for domestic supply, is more rigorously monitored and is generally controlled by treatment. The quality of natural waters can only be controlled by limiting discharges and is therefore subject to much wider variation. [See *Wastewater Treatment.*]

I. FRESH WATERS

A. Aquatic Ecology

Life in water is significantly different from that in terrestrial habitats. For aquatic organisms life is dominated by problems of obtaining sufficient oxygen and light or nutrients. Also in moving water there is the possibility of being washed away or in still water the chance of being buried under silt. Balanced against these difficulties are advantages such as greater thermal stability and the ready availability of water and dissolved salts.

The physical and chemical properties of warter are remarkable and have considerable significance in the study of aquatic ecology. Some knowledge of these peculiarities is a prerequisite for understanding water pollution. Although the chemical composition of water is conventionally expressed as H_2O, it behaves as $(H_2O)_n$ and in fact consists of continually changing branched chains of imperfect oxygen tetrahedra linked by hydrogen bridges. At low temperatures, near freezing, the tetrahedral arrangement is marked; as the temperature increases, there is less structural association so that the possibility of close random packing increases at the same time that increasing thermal agitation makes for a looser arrangement. The interaction of these two processes results in a contraction on melting and a further contraction up to 4°C. Beyond this temperature the thermal agitation becomes the dominant effect and water expands. The most important result of this phenomenon is in lakes and seas, where it is responsible for thermal stratification (see Section B), although an attenuated stratification also can be observed in some tropical rivers. Biological it is also important in ensuring the continuity of life beneath the ice in streams and lakes.

The presence of hydrogen bridges has other effects on the physical properties of water (*e.g.*, elevated boiling point and freezing point, high specific heat, and latent heat of sublimation and boiling), which ensure that water is liquid at most environmental temperatures and that it is extremely thermally stable. [See *Hydrogen Bonds*.]

The structure of water also affects its chemical properties, particularly in producing a high dielectric constant that makes it capable of dissolving almost any substance that can form an ionic solution. By contrast, substances that cannot ionize, such as oxygen and many organic compounds, are only sparingly soluble. The property of being such a good solvent has important biological consequences because as a result most natural waters contain enough inorganic material to support the growth of plant life.

Ecology is concerned with the relationship between organisms and their environment. The aquatic environment may be divided into a number of chemical and physical aspects, which are briefly discussed in this section.

1. Dissolved Gases

The concentration of dissolved gases depends upon the occurrence in the atmosphere, the solubility, and the rates of production and consumption within the aquatic habitat. The main dissolved gas of biological significance is oxygen, but carbon dioxide is also important. Oxygen has a very limited solubility in water, as shown in Fig. 1.1, and this fact limits both activity and abundance in many aquatic environments.

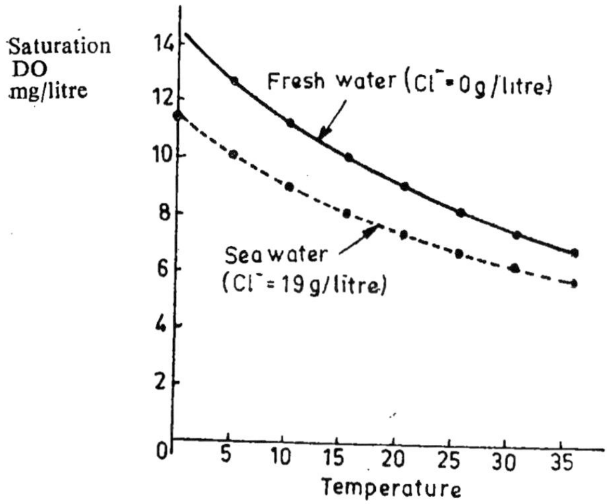

Fig. 1.1: Solubility of oxygen in water

2. *Dissolved Salts*

Since water is such a good solvent, it is inevitable that most natural waters and all wastewaters contain a variety of inorganic substances that are derived from the atmosphere, soil, rocks, and wastes. The ionic composition is important in determining growth rates of algae, invertebrates, and fish and in determining the water's suitability for domestic and industrial consumption (see Section 4, C). As total dissolved salts they fix the osmotic pressure, and aquatic environments may be divided into fresh water and seawater.

3. *Organics*

Organic matter enters aquatic environments by a variety of routes from terrestrial debris and the discharge of domestic and industrial wastes as well as from the waste products of aquatic organisms and the decay of their bodies after death. The level of dissolved and suspended organics plays a crucial role in determining whether an environment is dominated by primary producers or decomposers. Many organic substances are significant in other respects since they are often toxic or carcinogenic to humans or aquatic organisms.

4. Solar Radiation

Solar radiation is of crucial importance in providing the energy source for photosynthesis. Much of the solar radiation that falls on aquatic environments is lost by reflection (up to 50%). The subsequent penetration depends upon the incident light intensity, colour, and turbidity of the water, as shown in Fig. 1.2. The rate of reduction of light intensity (the extinction coefficient) is critical in determining the depth at which photosynthesis balances losses due to respiration (called the compensation depth). In clear mountain lakes and oceanic waters the compensation depth may be up to 50 m, whereas in waste stabilization ponds it may be as little as 0.2 m.

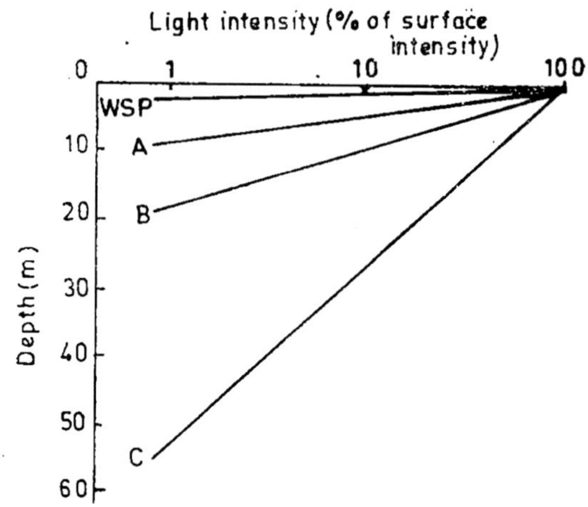

WSP=Waste Stabilization Pond
A=Eutrophic Lake
B=Turbid River
C=Oligotrophic Lake

Fig. 1.2: Light penetration in water

5. Temperature

Water temperature is very important to aquatic organisms since almost all of them are poikilothermic. The temperature requirements of different organisms vary, but the form of the response is similar and may be divided into zones of cold stress, optimum, and heat stress above which is the thermal death point. There is rarely any problem of cold death since natural waters or wastes do not often fall below zero, but freezing can cause severe rates of mortality. Within the feasible range for an organism, increase in temperature causes an increase in metabolic rate. The increase is usually 2-2.5 times for a 10°C rise, which can cause problems if the extra food and oxygen are not available.

On the basis of temperature patterns it is possible to distinguish three types of habitat: Cold and thermally stable habitats occur in the range 1°C to 6°C, such as the headwaters of a river, the hypolimnion of a temperate lake, and the deep waters in oceans. Warm and thermally stable environments occur in tropical rivers and lakes and surface waters in tropical oceans with temperatures varying between 20°C and 35°C. The third type of environment has temperature fluctuations large enough to be biologically significant (>10°C) and includes surface waters of lakes and seas and rivers in temperate regions.

6. Water Movements

Water movements are extremely important to aquatic organisms since with the exception of fish and aquatic mammals, aquatic organisms tend to be carried along by the movements. Freshwater aquatic habitats are divided into lentic and lotic, and the latter category can be subdivided on the rate of movement into different types of bed (silt, sand, gravel, boulders, and rock). In streams water movement is not uniform laterally or vertically. This has important consequences in providing different types of habitat and also in causing dispersion.

In lakes and seas movement of water takes the form of currents and waves. (See Section II, A for a discussion.)

B. Biotic Factors

Although the size and nature of an aquatic community are to a large extent determined by chemical and physical factors, their influence is considerably modified by the effect of biotic factors. The principal biotic factors are reviewed briefly in this section.

1. Trophic Relationships

The trophic structure of an aquatic community is illustrated diagrammatically in Fig. 1.3. Organisms on the same trophic level compete for food; those on successive levels have a preypredator relationship. Competition for food or sunlight or space is often the principal factor determining the degree of ecological success, with predation as an important secondary influence. Other types of trophic relationships are also significant, in particular the role of decomposers that use an alternative energy source and provide the basis for an alternative food chain and the role of parasitism in creating host-parasite relationships that can also affect ecological success.

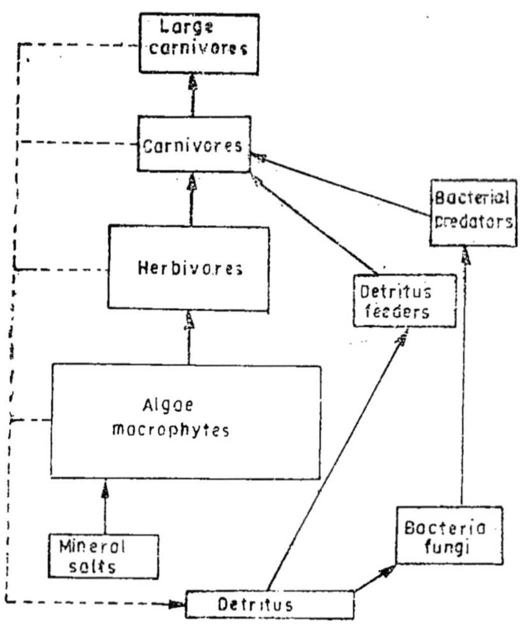

Fig. 1.3: Trophic structure of an aquate community

2. *Energy and Nutrient Flow*

The relationships with the energy sources enable organisms to be classified into various categories (producers; primary, secondary, or tertiary consumers; and primary or secondary transformers). As energy flows through the ecosystem, it is dissipated and only a limited amount of recycling takes place through dead remains and waste products contributing to the detritus. The balance in an ecosystem between the producer-consumer food chain and the transformer chain depends upon the relative energy inputs and is easily disturbed by pollution. (See Section I, C.)

The flow of nutrients in an aquatic ecosystem is much more conservative, often with a large proportion of elements such as nitrogen and phosphorus being recycled through the transformers back to the producers. Nutrient storage is a particular feature of lake ecosystems and may cause the long-term persistence of disturbances produced by nutrient enrichment. (See Section I, C.)

3. *Population Dynamics*

For individual species, factors such as growth rate, fecundity, and death rate play an important role in determining abundance. Organisms have different mortality and fecundity patterns with age, and for some particular stages in the life cycle may these patterns be critical.

C. Freshwater Pollution

The basic principles of aquatic ecology outlined in Section I apply equally to polluted sites. The following section discusses the ways in which pollution affects aquatic environments, in discussing these changes it is useful to classify them into the following categories: organic, toxic, solids, heat, inorganic nutrients, and oil. However, it should be appreciated that the ways in which these different forms of pollution affect a community also depend upon the nature of the habitat. In lotic waters the community is dominated by benthic organisms and it therefore more sensitive to the deposition of solids and organic enrichment. Lentic waters are dominated by plankton and are therefore more responsive to increase in inorganic nutrients.

A further complication arises from the interaction between different types of pollution. Although it is possible to find examples of waters affected by only one type of pollution, it is much more common to find environments subject to several types. [See *Pollution, environmental.*]

1. Organic Pollution

Pollution by organic matter is a complex phenomenon since it involves both the stimulation of the community by organic and subsequent inorganic enrichment and stress to the community due to deoxygenation and alteration of the bed by slime growths. The chemical changes and their biological consequences are represented. The primary effect of organic enrichment on the microbial community is to stimulate the growth of heterotrophic bacteria. Some of these, especially the filamentous forms, grow in association with fungi and stalked ciliate Protozoa to form an attached growth called sewage fungus. This forms a characteristic and visible zone just downstream of significant discharges of organic waste, particularly where the discharge contains settleable organic solids that coat the bed. The result of this intense microbiological activity is to deplete the oxygen more rapidly than it can be replaced by reaeration; this leads to an oxygen sag. The extent of the sag governs the changes in the invertebrate community and fish as shown in Table 1.2 and is the principal cause of biological changes downstream of an organic discharge. The oxygen balance has been represented in terms of two competing processes, deoxygenation and reaeration, but in practice it is more complicated, with other processes such as photosynthesis and benthic respiration playing a dominant role. Mathematical modelling has therefore been extensively employed in quantifying the relationship between the amount of organic matter discharged [as measured by biochemical oxygen demand] and the resulting dissolved oxygen.

TABLE 1.2

Minimum Tolerance Levels for Dissolved Oxygen

Organisms	*Minimum level of dissolved oxygen*
Salmon, trout, mayflies, stoneflies	6-7 mg/litre
Perch, caddisflies, crayfish	5-6 mg/litre
Roach, tench, shrimps	3-5 mg/litre
Carp, eels, midges	1-3 mg/litre

The other results of the microbiological activity is the conversion of organic nitrogen and phosphorus into inorganic forms. Provided that other conditions [light intensity dissolved oxygen and stream bed] are suitable, production of inorganic nutrients will lead to increased growths of algae, especially attached forms such as Cladophora and Vancheria. Further downstream the diversity of algae increases, but this is accompanied by a decrease in algal biomass as the excess nitrogen and phosphorus are used up. Rooted plants and mosses are generally eliminated or much reduced by organic discharges due mainly to being smothered by silt or sewage fungus.

Organic enrichment and oxygen depletion stimulate the growth of Protozoa and detritus-feeding Metazoa at the expense of other invertebrates, especially if there is solids settlement. Where the stream becomes anaerobic, only the air-breathing animals survive. In situations with low DO (1-2 mg/litre) and a sludge blanket, the fauna is dominated by Tubificid worms. At slightly higher DO levels (2-3 mg/litre) chironomid larvae compete more successfully for the organic matter, and at even higher levels (3-5 mg/litre) other organisms like Asellus and the Gammarus succeed together with leeches, molluscs, and other fly larvae. This is followed by a gradual return to the normal clean-water fauna as the effects of organic enrichment die away.

In the zone affected by the organic enrichment there is an increase in the total biomass of the benthic fauna due to the increased food supply. Those organisms that can tolerate the conditions may be present in vast numbers, partly due to the extra food and partly due to reduced competition and predation. Where the concentration of organic matter from an effluent is low (BOD$<$ 4 mg/litre), the effects of organic enrichment are observable without the attendant deoxygenation and deposition of solids. In such cases there is a slight change in composition of the benthic fauna in favour of detritus feeders together with an increase in overall biomass. Where the effects of organic pollution are severe, pollution can lead to the elimination of fish populations from long reaches of a stream. Less severe pollution will enable the more tolerant fish to survive, and since the food supply is ample will lead to increased numbers. For the more sensitive species of fish to survive organic discharges, the reduction in DO must be limited to only 2 to 3 mg/litre, as indicated in Table 1.2. The standards set for controlling the effects of organic pollution are discussed in Section I, D.

2. *Toxic Pollution*

Apart from ammonia, domestic wastes do not contain any material that causes poisoning of aquatic organisms, but there are a vast number of toxic substances in industrial wastewaters. Toxic substances may be in solution and enter organisms such as algae, macrophytes, and invertebrates by diffusion through external surfaces or via the internal surfaces of fish, where the gills are the main portal of entry. The other mode of entry is through the food when the toxic substance is released during digestion. For some poisons there is a mechanism of excretion and/or breakdown so that a balance between uptake and elimination is possible, but for other toxins such mechnisms are ineffective and accumulation takes place. Toxins are generally divided into acute and chronic. Acute toxins exert their action over a short period, usually less than a week, whereas chronic poisons continue to act over a period of months or even years. Some poisons have a particular affinity for certain organs such as the liver, but others accumulate more generally. Because of these physiological complications, it is difficult to generalize about toxic action, but Table 1.3 indicates some general classes of poisons. [See *Water Pollution and Toxicology*.]

TABLE 1.3

Mode of Action of Some Common Aquatic Toxins

Class of poison	*Mode of action*	*Examples*
Heavy metals	Precipitation of proteins in gill secretions and anemia	Lead, mercury
Respiratory depressants	Inactivation of enzyme systems related to respiration	Cyanide, sulphide
Inorganic acids and alkalis	Corrosive poisons that attack soft tissues	Sulphuric acid, sodium hydroxide
Phenolic substances	Corrosive poisons that also attack central nervous system	Phenol, acridine
Synthetic detergents	Cause respiratory distress by attacking gill surface	ABS, lissapol
Insecticides and herbicides	Disturb the action of the central nervous system	DDT, malathion
Amonia	Upsets water balance by increasing permeability	

Since it is not possible to predict a toxin's effects from a consideration of its chemical composition, it is necessary to establish toxic effects experimentally. The test procedure is a form of bioassay in which test animals are exposed to a range of concentrations of toxin for periods up to 10 days. There are many practical difficulties in standardizing the procedure (such as choice of test animal), but the main difficulty is interpretation. As shown in Fig. 1.4, the test result can be used to estimate the safe concentration by plotting the median survival times in each tank against concentration and from the asymptote determining the concentration below which survival is independent of the toxin (*i.e.*, the safe concentration). Alternatively, the results can be used to calculate the concentration that poisons 50 per cent after all the toxic action has been exerted (incipient TL_m). This is a useful basis for comparing the relative toxicities of different poisons and can be used for estimating the safe concentrations by using the equation

safe concentration

$$= \text{incipient } TL_m \times \text{application factor}$$

but the application factor is difficult to establish.

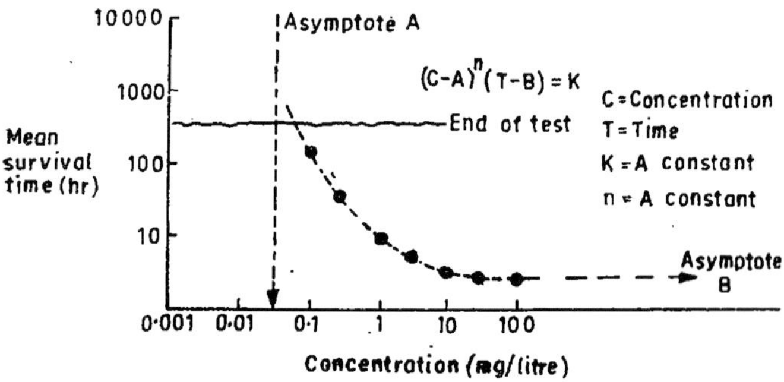

Fig.1.4: Interpretation of the results of an acute toxicity test

For chronic toxins all toxicity is not exerted within the 10-day period, and this type of test is therefore inappropriate, although a 96-hr test is sometimes used in conjunction with an application factor of 0.01 to estimate the safe concentration. Long-term growth tests are a better basis for setting safe standards for chronic toxins.

In polluted waters mixtures of poisons are commonly encountered, and it is therefore useful to be able to determine the toxicity of mixtures. It is usually assumed that the toxicity is additive and that the combined toxicity is the sum of the individual toxic effects expressed in toxic units. These are calculated as

$$\text{toxic units} = \frac{\text{actual concentration} \; - \text{threshold concentration}}{\text{incipient } TL_m}$$

This approach has been shown to give good agreement in over 80 per cent of cases and near agreement in a further 10 per cent. In the remainder chemical reactions among toxins tend to reduce the toxicity.

Fluctuating concentrations also present a problem in calculating toxic effects. Various approaches have been tried, of which the most useful is the idea of mortification units. These are calculated by multiplying the time of exposure by the concentration of the toxin. Mortification units have been shown to be a good basis for comparing exposures.

Two other aspects of the biological response add further complications to toxic pollution. The first is acclimatization. This is a general property of all types of aquatic organisms and describes their increased resistance to a poison on second or subsequent exposures. Acquired tolerance to toxins is even more important in field situations, where it develops in populations via natural selection.

The second aspect is limited to fish and some large invertebrates. These more mobile species are sometimes capable of detecting poisons and may be able to avoid them by moving into tributaries or backwaters. The ability to detect poisons seems to be limited to a relatively small range of compounds, mainly corrosive substances.

From the preceding discussion it is apparent the toxic substances can cause severe pollution and may damage aquatic environments in many subtle ways that are difficult to quantify. Assessment of toxic pollution and standards are discussed later.

3. *Solids Pollution*

Particulates are discharged into water bodies through various processes, including erosion of soil and rock through land run-off, leaf fall, discharge of sewage, discharge of industrial wastes, and discharge of urban drainage. The effect of these particles on aquatic life depends primarily on whether they remain in suspension and to a lesser extent on whether they are organic or inorganic. Where solids remain in suspension, they reduce the photosynthetic activity by increasing the turbidity, and at high concentrations (>300-400 mg/litre) they can have a direct abrasive action on soft tissues such as the gills of fish. But for concentrations of suspended solids of less than 100 mg/litre, there is no evidence of any damage to aquatic communities unless the solids are readily degradable organics and cause other problems through deoxygenation.

Where solids settle they can cause severe damage to aquatic communities in very small amounts. This is especially the case when the deposition takes place in the headwaters of rivers used for breeding grounds by salmonid fish. But many invertebrates are also badly affected when the interstices of eroding gravel are blocked by fine solids, and a layer of only 10 to 20 mm is required to completely eliminate some species by increased stream drift.

The nature of the depositing solids does affect the extent of the changes since organic solids are particularly damaging. They often lead to totally anaerobic conditions that eliminate all but air-breathing invertebrates and bacteria.

The main problem in assessing the effect of the discharge of solids into water bodies is in predicting their hydraulic behaviour, particularly in streams where conditions can change from deposition to erosion with changing river flow. Setting standards for solids of therefore particularly difficult (see Section D).

4. Heated Discharges

Increasing the temperature of aquatic environments can damage the community in various ways, notably by

1. Lowering the solubility of dissolved oxygen,
2. Decreasing the availability of oxygen by stimulating bacterial activity (the increase in activity is approximately 10 per cent per degree Celsius),

3. Increasing the requirements for both food and oxygen of aquatic organisms (the increase in metabolic rate varies for different organisms but can be up to 250 per cent per 10°C increase),
4. Disturbing the reproduction pattern, particularly of fish, so that males or females are mature at the wrong time or the eggs hatch too soon, and
5. Causing physiological distress (for most aquatic organisms there is a very small range between optimum and maximum temperatures).

Heated discharges arise from a variety of industrial operations such as smelting and coke production, but they mainly come from thermal power stations. Because of the fluctuating nature of the demand for electricity, there are often marked seasonal and diurnal fluctuations in the quantities of heat discharged. This creates a similar problem to that caused by fluctuating concentrations of toxins (see Section I, C) and is usually handled in a similar manner by the use of mortification units.

Acclimatization to heat is also similar to acclimatization to toxins. This is illustrated in Fig. 1.5 (See also Section D.)

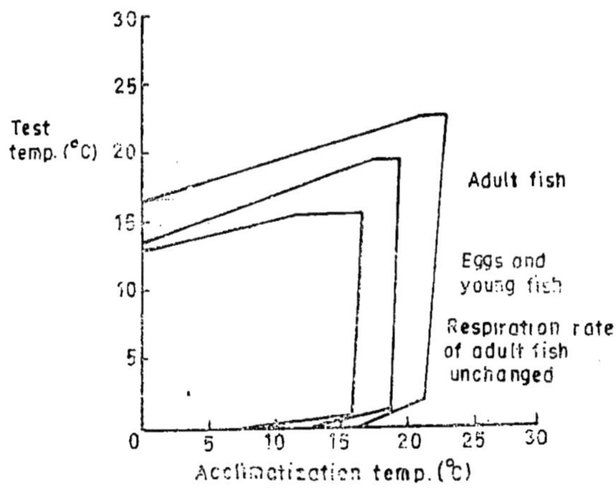

Fig. 1.5: Acclimatization to heat by an aquatic organism

5. Eutrophication

The concentration of inorganic forms of nitrogen and phosphorus (and sometimes silicon) often restricts the amount of algal growth, especially in lentic waters that are particularly suitable for planktonic organisms. Ammoniacal nitrogen is toxic to some aquatic organisms (see Section I, C), and concentrations of nitric nitrogen above 10 mg/litre are associated with increased incidence of methaemaglobinemia (see Section III, C), but the main effect is eutrophication.

The general effect of inorganic nutrient release in rivers has already been discussed in relation to organic discharges. Where inorganic materials are discharged directly into streams, the same effects appear but may be in a more severe form. Large growths of Cladophora (and other filamentous algae may make the river unsightly and in association with rooted vegetation may alter the nature of the bed and cause siltation. But the chief concern of nutrient addition to streams is where the water is subsequently impounded. Nutrient enrichment in lakes and reservoirs takes various forms, depending on whether the water body is used for recreation or water supply. Recreational problems are caused by algal growth causing diurnal fluctuations in pH, alkalinity, and dissolved oxygen and through causing increased turbidity and in extreme cases producing floating mats of algae. The problems caused in abstracted waters are mainly due to increased levels of turbidity plus other problems of taste and odour. These are discussed in more details in Section III, C.

In considering the significance of nutrient discharges it is important to have a grasp of the levels at which nitrogen and phosphorus affect algal growth. These are summarized in Table 1.4 as average levels for minimum and maximum growth. Individuals species obviously very considerably in their requirements. For example, blue-green algae are capable of utilizing dissolved nitrogen gas and are therefore independent of ammoniacal and nitric forms. But the figures do show the higher nitrogen requirement compared with phosphorus and the much higher nutrient requirements of the benthic algae.

Nutrient enrichment of streams and lakes is due to contributions from many sources. These are summarized in Table 1.5.

TABLE 1.4

Approximate Requirement of Nitrogen and Phosphorus in Relation to Algal Growth (All Levels in μg/litre)

Nutrient	*Plankton*	*Benthic*
Minimum level of nitrogen	10	500
Saturation level of nitrogen	1000	5000
Minimum level of phosphorus	1	100
Saturation level of phosphorus	100	1500

TABLE 1.5

Sources of Nitrogen and Phosphorus Discharges

Source	*Nutrient load*	
	Nitrogen	*Phosphorus*
Domestic waste	9 g/person day	2 g/person day
Industrial wastes	0 up to fertilizer waste	Usually low except detergent-rich wastes
Urban runoff	2-6 kg/ha yr	1-2.5 kg/ha yr
Arable runoff	10-20 kg/ha yr	0.5-1 kg/ha yr
Pasture runoff	6-10 kg/ha yr	0.1-05 kg/ha yr

The pattern of nitrogen and phosphorus levels differs geographically. In Europe and the United States, phosphorus is the main limiting nutrient in freshwater systems and its main source is detergents. Point sources contribute up to 80 per cent of the total phosphorus budget. In tropical areas of Africa, South America, and Asia, soils are deficient in nitrogen and this is the main limiting nutrient in most water bodies. This is in marked contrast to countries such as the Uuited Kingtom, where lower rainfall combined with extensive use of nitrogenous fertilizers lead to significant runoff of nitrogen from farmland, which is responsible for over 60 per cent of the total nitrogen budget in rivers.

Since the consequences of eutrophication are more serious, greater attention has been paid to the connection between nutrient loads and the levels of algae. The relationship between nutrient inflow to a reservoir and the nutrient concentration in the reservoir water is somewhat complex. In general terms the degree to which the nutrient content of the reservoir will approach that of inflows is affected by retention time, number of algae present, and their rate of growth. In temperate reservoirs with seasonal growth, there is some correlation between winter nutrient levels and the algal concentrations in the following summer, as shown in Fig. 1.6.

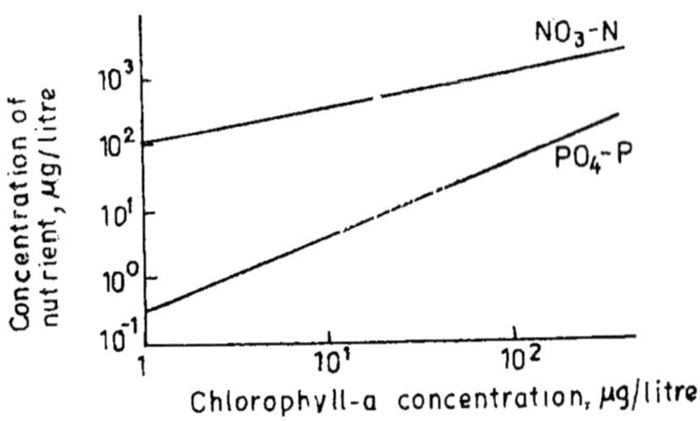

Fig. 1.6: Correlation between winter nutrient levels and algal concentrations in summer. [Reprinted with permission from Lund, J, W. G. (1971). Eutrophication. In "The Scientific Management of Animal and plant Communities for Conservation." Symp, Br. Ecol. Soc. (E. Duffy and A.S. Watts, eds.). Blackwell London.]

A more generalized approach is to assess the potential for eutrophication in terms of the nutrient loading, as illustrated in Table 1.6. This has been shown to apply to many European lakes and has since been found to be applicable in other parts of the world.

TABLE 1.6

Permissible Loadings of Nutrients for Lakes and Reservoirs

Mean depth	*Permissible loading*		*Dangerous loading*	
	N	*P*	*N*	*P*
5	1.0	0.07	2.0	0.13
10	1.5	0.10	3.0	0.20
50	4.0	0.25	8.0	0.50
100	6.0	0.40	12.0	0.80

6. *Oil Pollution*

Oil pollution in fresh water is generally an aesthetic rather than a biological problem. Small spillages of oil can create visible surface films over a wide area, and this is a common sourse of complaint to water pollution control organizations.

Where large quantities of oil are discharged, oxygen depletion and smothering action may damage the aquatic community. But the effects are short-lived and the community usually recovers rapidly.

The effects of oil pollution are more frequently encountered in the marine environment (see Section II, B, 4).

D. Assessment of Freshwater Pollution

Water pollution is a biological phenomenon that ought to be assessed biologically. There are many advantages in using biological assessment, notably that benthic organisms intergrate the numerous environmental influence and give a direct answer to whether a particular combination is suitable. Second, the effects of pollution are integrated over time; especially with the longer-lived macroinvertebrates, the presence of mature individuals can indicate appropriate conditions over the past 6 to 12 months. Also biological data clearly show sublethal as well as lethal effects. The former are extremely difficult to discern from purely chemical observation.

Despite these obvious advantages, most water pollution assessment is based on chemical monitoring. This is, to some extent, a reflection of the ease with which collection and analysis can be carried out, but it is mainly due to the quantitative nature of the answer that is obtained. This may best be illustrated by reference to the examples of chemical and biological classification of rivers in Tables 1.7 and 1.8. Where a standard of 4 mg/litre is adopted for BOD, an analytical result will give a quantitative indication of the amount additional treatment required However, a result from a biological survey classified as class III gives no real indication of the BOD removal needed to bring the river up to class VII or VIII.

One further problem with biological assessment is the difficulty in devising a universal system. Most systems rely on changes in the community due to some or all of the following factors: diversity, range of species present; spectrum, type of species present; and abundance, number of individuals of each species present. But the changes in these parameters due to pollution are superimposed on the changes due to other edaphic factors, and so each system tends to work best in the area for which it was devised and cannot easily be applied to other regions without modification. Some combination of chemical and biological assessment is therefore required for satisfactory monitoring of water pollution.

Special problems in assessment are posed by suspended solids and heated discharges. The control of suspended solids has to be exercised on an individual basis for each stream situation since the effects depend upon the local hydraulics.

The effect of heated discharges is also site dependent. [See *Environmental Data Acquisition.*]

Suggested standards for these are summarized in Table 1.9.

TABLE 1.7

Chemical Classification of River Water Quality in the United Kingdom

River class	*Quality criteria*	*Current and potential use*
1A	DO > 80% saturation; BOD <3 mg/litre; ammonia < 0.4 mg/litre; nontoxic to fish	Water of high quality suitable for game fishery and with a high amenity value
1B	DO > 60% saturation; BOD<5 mg/litre; ammonia < 0.9 mg/litre; nontoxic to fish	Water of less high quality but suitable for substantially the same purposes
2	DO>10% saturation; BOD<9 mg/litre; nontoxic to fish	Water supporting coarse fishery and of moderate amenity value
3	DO>10% saturation; BOD<17 mg/litre	Fish only sporadically present but water unlikely to cause a nuisance
4	Inferior to class 3	Waters that are grossly polluted and liable to cause a nuisance

II. MARINE WATERS

A. Marine Ecology

Although the sea is an aquatic environment that shares many of the characteristics of freshwater environments, there are some important differences.

TABLE 1.8

Present Biotic Index

Indicator group	*Diversity of indicator group*	*Number of other groups present*				
		0-1	*2-5*	*6-10*	*11-15*	*16+*
Plecoptera nymphs present	>1 species	—	VII	VIII	IX	X
	Only 1 species	—	VI	VIII	VIII	IX
Ephemeroptera nymphs present	>1 species	—	VI	VII	VIII	IX
	Only 1 species	—	V	VI	VII	VIII
Tricoptera larvae present	>1 species	—	V	VI	VII	VIII
	Only 1 species	IV	IV	V	VI	VII
Gammarus present	All above species present	III	IV	V	VI	VII
Asellus present	All above species present	II	III	IV	V	VI
Tubificid worm and/or Chironomid larvae present	All above species absent	I	II	III	IV	—
All above types absent	Only organisms not needing DO present	O	I	II		

TABLE 1.9

Suggested Standards for Pollutants

Pollutant	*Guideline*
Inert solids	<80 mg/litre if solids remain in suspension
	<10 mm deposit on bed if solids settle (especially in eroding environments)
Organic solids	<6 mg/litre as BOD if solids remain in suspension
	<0.1 g O_2/m^2 hr as sediment oxygen demand
Heated discharge	<8°C temperature rise in streams at any time
	<2°C increase in summer maximum for Salmonid fish
	<6°C increase in summer maximum for coarse fish

1. Osmotic Pressure

The salt concentration in seawater (34-37 g/litre)it too high for the majority of organisms adapted to fresh water. Very few organisms can tolerate both freshwater and marine conditions, although there are important exceptions, such as salmon and eels. The abrupt biological transition caused by increased salinity can be abserved in estuaries, as shown in Fig. 1.7.

2. Wave Action

A wave is composed of a cone of rotational movements in which velocities may be sufficient to cause scouring of solids from the bed. In most seas the waves generated at the surface can only resuspend solids up to a depth of about 50 m. Any solids that settle in deeper waters remin undisturbed, but at lower depths they will be resuspended. Wave action is also very important to littoral communities.

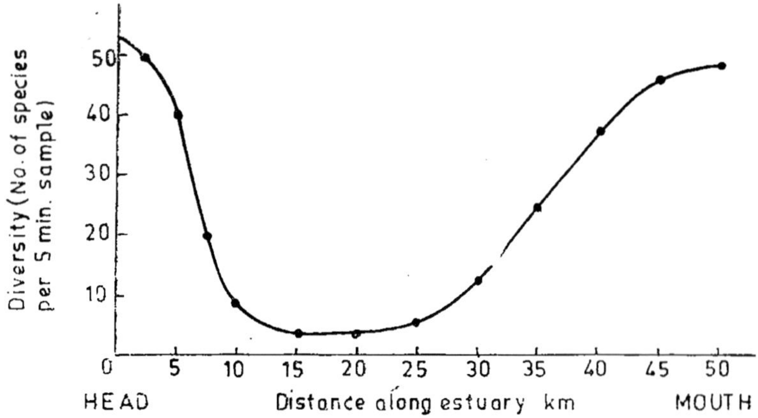

Fig. 1.7: Diversity in an unpolluted estuary

3. Currents

There are three main types of currents that are important: tidal, wind induced, and oceanic. Tidal currents are responsible for much of the advection and dispersion that take place near coastlines. The majority of wastes enter the sea in the zone from estuaries or outfalls and the mixing and movements caused by tidal currents are therefore critical in determining the pattern of concentration around the point of discharge.

Wind-induced currents are limited to the top 1 to 2 m. Within the top metre velocities may be 5 per cent to 10 per cent of the wind velocity and cause rapid translocation of bacteria, oil, and other surface pollutants back onto a beach. They are of particular interest in the design of marine outfalls.

Oceanic currents are caused by convection in tropical regions, which induced surface movements toward the pole and deep-water return currents. These are relatively slow (a few kilometres per day) but over a long period of time are responsible for the global distribution of polutants.

4. Dilution

The sea has an almost unlimited capacity for dilution. provided that polluting materials are rapidly disperssed, the resulting concentations are below toxic thresholds and the rates of decomposition of organic matter make tittle impact on the dissolved oxygen concentration. A corollary of the great dilution capacity is an almost infinite retention time, and any pollutants discharged to the sea that do not degrade will remain there almost indefinitely. This gives rise to the possibility of reconcentration through biological agencies (food chains) or geochemical processes (adsorption or sediments) or both. Reconcentration may cause persistent pollutants to exceed toxic thresholds, particluarly in predators near the top of a food chain.

5. *Light*

The light and temperature pattern in the sea is similar to that in lakes except that the much greater depths dictate that the vast bulk of the seas (*i.e.*, below 100 m) are permanently cold($<2°C$) and dark.

6. *Nutrients*

Seawater is deficient in both nitrogen and phosphorus, the background concentrations being 0.2 μg/litre and 0.02 μg/litre, respectively. This imposes a severe limitation on primary production.

B. Pollution

The special physical and chemical features of marine environments discussed in the preceding section are reflected biologically in the community structure and composition. They are also reflected in the effects of pollutants. Due to the rapid rate of dispersion (caused by density differences and current shear) and the enormous dilution available, the polluting action of BOD, suspended solids, and acute toxins cause only localized problems. The important types of marine pollution are toxic pollution, slick formation, bacterial pollution of bathing beaches, and oil pollution.

1. Pollution by Persistent Toxins

The two groups of substances involved are metals (mainly mercury, cadmium, and lead, although many other metallic elements may be involved) and chlorinated organics (mainly pesticides and herbicides such as DDT and the eldrins along with chlorinated organics used for other purposes, such as PCBs). Both groups of substances are extremely susceptible to adsorption onto solids and are rapidly removed from solution in the sea. They therefore accumulate in marine sediments (particularly in onshore areas) from where they may remobilize due to chemical action or may enter marine food chains by the feeding of benthic invertebrates.

The pollutants are notable for the extremely low threshold of their toxic effects (*e.g.*, cadmium 10 μg/litre and dieldrin 10 ng/litre). Concern over this type of pollution has become severe due to the worldwide nature of the problem and its irreversible natutre. Recent legislation has attempted to limit any further discharge of persistent pollutants into the sea.

2. Surface Slick Formation

Discharges from submarine outfalls may cause an undesirable surface appearance if they do not achieve sufficient dilution before reaching the surface. To avoid slick foration it may be necessary to extend pipelines further offshore into deeper water to or pretreat the waste.

3. Microbiological Contamination of Bathing Waters

The relationship between the incidence of intestional disease in batthers and the microbiological quality of bathing water is very complex, but enough evidence of some causal connection has accumulated to prompt the introduction of standards. These standards generally specify the quality requirements for bathing waters in terms of coliforms or fecal streptococci, but recent evidence has suggested that the main risk to health is due to enteric viruses. Therefore, viral standards may soon be imposed.

The calcultation of standards and the engineering methods for achieving them are complicated by the differential die-off behaviour of pathogens and indicators in the sea

The other health hazard posed by discharge of bacteria to the sea is from shellfish. Because of their filter-feeding methods, these organisms can concentrate pathogenic microorganisms (and heavy metals) up to dangerous levels. Therefore, protection of shellfish grounds is required.

4. Oil Pollution

Marine environments are particularly susceptible to pollution by oil from accidents, spillages, and deliberate discharges of bilge waters associated with the global transport by sea. Very small quantities of oil are needed to produce a visible slick around 0.0001-mm-thick layer). Thicker layers can cause biological damages especially to seabirds and littoral communities through toxic action and by boating the organisms and preventing respiration, feeding, and photosynthesis. The toxicity is inversely correlated with the molecular weight, and as the lighter, more volatile fractions evaporate, the oil becomes less toxic (96-hr TL*m* going from around 10 mg/litre to over 1000 mg/litre). With the possible exception of seabirds, the effects of oil pollution do not have long-term effects on marine populations. The fate of oil in the sea depends to some extent on the oil's molecular weight. The lower-molecular-weight fraction as well as evaporating is more easily emulsified, and the resulting droplets are rapidly degraded by phytoplankton and bacteria. The heavier fractions tend to settle and degrade much more slowly, but biological effects even from severe pollution do not usually persist for more than 1 or 2 years. The main reason for control measures is, therefore, to avoid damage to tourist areas. Control is by burning, accelerated sinking, or increased dispersion, although containment and recovery may occasionally be possible.

III. ABSTRACTED WATERS

A. Sources

Water is abstracted from various natural sources including rainwater, streams, lakes, underground water, and even the sea. In judging the suitability of a source, the following considerations apply.

1. Reliable Yield

Sources such as rainwater are essentially temporary in nature and therefore cannot be relied upon to give a continuous supply. This applies to a lesser degree to streams whose flow may be reduced during drought, and even lakes and underground sources are not completely reliable; only the sea has sufficient storage capacity to guarantee any demand. Water sources are therefore assessed in terms of the yield that can be abstracted for a given probability of failure.

2. Quality

Water may be abstracted for many purposes, some of which have no particular requirements for quality such as hydro-electric power generation or fire fighting. But for most domestic, industrial, and agricultural purposes, quality considerations are important and the suitability of a source is largely judged on its quality. It is generally possible to improve the quality of abstracted waters by appropriate treatment, but where large flows are involved such as in irrigation this may not be economically feasible.

3. Location and Conveyancing

In many countries the main sources of population and industry are situated away from the wetter regions, and so a large part of the cost of developing water sources may be due to conveyance, especially where insufficient head is available to allow the water to flow by gravity. The location may also involve legal problems in access or protection of the catchment, which may be in conflict with other requirements.

Where alternative sources are available, they are evaluated according to the criteria just described in relation to the anticipated demand, the quality required, the reliability required, and the capital and recurrent revenue available. In general terms rainwater is rarely a reliable source except for individual households, and the cost of treating seawater is prohibitive where other sources are available. Upland surface sources are preferable in terms of quality. Groundwaters, especially those deeper than 30 to 150 m, are usually the best for reliability and quality but can be expensive for pumping unless they are under artesian pressure. Shallow groundwater sources are liable to contamination, and groundwaters deeper than 200 to 300 m are often too high in total dissolved solids.

B. Collection and Storage

The quality of water in any natural system is continually changing. This applies to water that is being collected and stored for abstraction, and so it is sensible to consider these as part of the water treatment process.

In the case of surface waters it has long been recognized that protection of the catchment is essential in minimizing the cost of obtaining the required water quality. But the emphasis on catchment protection has changed with improved technology of treatment and increasing pressures on catchment areas for recreation. The main priority in catchment management is concentrated on reducing nutrient loads entering reservoirs so as to minimize algal growths. The second important aspect of catchment management is concerned with prevention or reduction of soil erosion, since this not only reduces the storage capacity, but also can impair the water quality due to increased BOD or suspended solids. The importance previously attached to avoiding bacterial contamination of catchments has been somewhat reduced. Although bacterial contamination is still regarded as undesirable, it is appreciated that the complete prohibition of recreation and agriculture are unnecessary since the technology for bacterial reduction is reliable and relatively inexpensive.

The need for storage for a surface source can have a major influence on the quality of the water. The water quality may improve due to sedimentation, oxidation of organic matter, and die-off of bacteria and other pathogens. But the water quality may also deteriorate mainly due to algal growth. Factors such as thermal stratification and eutrophication largely control the changes in quality during storage and need to be carefully considered during the design and construction of a reservoir. [See *Water Resource Systems*.]

Once a lake or reservoir is used for storage, algal growths need to be monitored and if necessary controlled. Techniques for destratification by induced circulation are available mainly by the use of compressed air and submerged jets. These same techniques help to reduce the rate of algal growth and prevent deoxygenation of the hypolimnion with the consequent problems of sulphide production and increased levels of manganese and iron. Alternatively, algal growths may be controlled by the use of algicides. Various compounds have been found that successfully inhibit the growth of algae, notably copper sulphate and 2:4 dinaphthoquinone (2:4D), but these are expensive, may create secondary toxicity problems, and provide only a temporary amelioration since the algae will grow again the following year.

Where it is impossible or uneconomical to control the growth of algae, the water leaving the reservoir or lake will require additional treatment to remove not only the algae, but also their metabolic products, which can give rise to unpleasant tastes and odours. Also, chlorination of these metabolic products can lead to a risk of the formation of carcinogens if they are not removed.

Surface storage of water in tropical climates can have additional problems due to the growth of aquatic macrophytes. The difficulties caused by these large macroscopic growths chiefly affect fishing, navigation, and recreational uses, but such growths also affect water treatment plants adversely by causing blockage of intakes, disruption of sedimentation, and other problems, as well as by increasing the organic debris in the water.

Storage of water underground is generally free from the problems that beset surface storage, particularly where the groundwater is in a confined aquifer. However, problems of chemical contamination can arise due to leaching from waste tips, discharge from septic tanks, or spraying of pesticides and herbicides. Where the aquifer is near the surface or contains fissures it is also possible for microorganisms to gain access.

Various chemical and biological processes occur in the stored water mainly because of bacteriological activity, which leads to a decrease in organic content, a decrease in dissolved oxygen, and an increase in carbon dioxide. The increase in CO_2 may cause a fall in the pH and where the aquifer is calcareous can lead to an increase in hardness. Other chemical changes occur due to contact with minerals in the aquifer. Often the rock strata can behave as an ion exchange medium that alters the cation composition. Metal complexes may be formed owing to the high carbonic content that brings previously insoluble metals such as uranium into solution as complex hydrated carbonates.

The main dangers to the quality of water stored underground come from saline intrusion and contamination from leachates. Saline intrusion can occur when the drawdown of the water table allows seawater to enter the aquifer. This problem is usually confined to coastal areas but can place severe constraint on the yield from the aquifer. Contamination from leachates is nearly always chemical in nature since the movement of bacteria or other pathogens is very limited underground unless large fissures are present. The material that causes most difficulties is nitrate since this is usually used as a fertilizer, is readily soluble, and is not easily removed during storage underground. The other agricultural sources of pollution are pesticides and herbicides, particularly chlorinated organics. Other sources are deposits of domestic or industrial sludge and solid wastes, which may yield metals or other toxins.

C. Quality Requirements

The quality of a water is judged in relation to its use. This principle applies especially where the water quality is being changed by treatment since unnecessary expenditure may be incurred. The quality requirements for domestic supplies, for industry, and for agriculture will therefore be considered separately.

Domestic water supplies must satisfy three criteria:

1. they must be free from injurious chemicals,
2. they must be free from harmful pathogens, and
3. they must be aesthetically satisfactory.

The list of potential toxins in water is virtually infinite since even a substance like common salt becomes harmful if present in sufficient concentration. Table 1.10 summarizes the limits as suggested by World Health Organization to some common toxins. It must be appreciated that the guidelines given in Table 1.10 refer to the quality of water as supplied to the consumer. It is perfectly normal to take water that does not conform to these requirements and then alter the quality by appropriate treatment.

The processes normally used in water treatment are a combination of flocculation, sedimentation, filtration, and disinfection. With the exception of disinfection, they are designed for solid removal. Pollutants that are in suspension are therefore removed effectively as are those that are absorbed onto solids (*e.g.*, metals), but pollutants that are in true solution such as nitrates are not readily removed by conventional treatment and require special techniques.

Much of the natural colouring meterial in surface waters is humic acid derived from decomposition of plant debris. This is present as negatively charged colloids and is effectively removed by the addition of a coagulant (aluminium or iron salts), which destabilizes the colloid and forms a sludge blanket that helps with sediment removal. Adjustment of pH and alkalinity is often required to keep these processes at maximum efficiency.

TABLE 1.10

Inorganic and Organic Constituents of Health Significance*

Constituent	*Unit*	*Guideline value*	*Remark*
1	*2*	*3*	*4*
Arsenic	mg/litre	0.05	
Cadmium	mg/litre	0.005	
Chromium	mg/litre	0.05	
Cyanide	ma/litre	0.1	
Fluoride	mg/litre	1.5	Level varies due to climatic conditions
Lead	mg/litre	0.5	
Mercury	mg/litre	0.001	
Nitrate	mg/litre	10	
Selenium	mg/litre	0.01	
Aldrin or dieldrin	μg/litre	0.03	
Benzene	μg/litre	10	Based on mathematical model, not verified
Carbon tetrachloride	μg/litre	3	Tentative guideline
Chlordane	μg/litre	0.3	
Chloroform, unverified	μg/litre	30	Based on unverified model
2.4D	μg/litre	104	
DDT	μg/litre	1	
1.2-dichloroethane	μg/litre	10	Based on unverified model
1.1-dichloroethane	μg/litre	0.3	Based on unverified model
Heptochlor	μg/litre	0.1	

(*Contd.*)

1	*2*	*3*	*4*
Hexachlorobenzene	μg/litre	0.01	Based on unverified model
Lindane	μg/litre	3	
Methoxychlor	μg/litre	30	
Pentachlorophenol	μg/litre	10	
Tetrachloroethane	μg/litre	10	Based on unverified model
Trichloroethane	μg/litre	30	Based on unverified model
2,4,6-Trichlorophenol	μg/litre	30	Based on unverified model

*Summarized from WHO Guidelines for drinking Water Quality, 1984.

The requirement to be free from harmful pathogens is to avoid the risk of spreading disease, since waterborne infections have been the cause of much morbidity and mortality. The microbiological criteria used in assessing the quality of drinking water are summarized in Table 1.11. One important feature of the microbiological guidelines is their reliance on indicator bacteria. It is not normal practice in water examination to search for pathogens such as Vibrio cholerae of Salmonella typhosa for the following reasons.

1. The aim of the examination is not to establish the presence of pathogens, but to search for evidence of fecal contamination.
2. The pathogen content of sewage is small and extremely variable, whereas the indicator organisms give a much more reliable quantitative assessment of the degree of fecal contamination.
3. Techniques for the enumeration of pathogens are not as reliable and quantitative as those for coliforms and other indicators.

TABLE 1.11

Microbiological Guidelines for Drinking Water Quality as Suggested by WHO (1984)

Types of supply	*Indicator bacteria*	*Guideline value*	*Remark*
Piped supply of treated water entering distribution system	Fecal coliforms	0/100 ml	Turbidity <1 NTU for disinfection with chlorine
	Total coliforms	0/100 ml	pH preferably <8.0; free chlorine residual of 0.2-05 mg/litre after minimum contact time of 30 min
Piped supply of untreated water entering distribution system	Fecal coliforms	0/100 ml	In 98 per cent of samples per year
	Total coliforms	0/100 ml	In remaining 2 per cent, which
	Total coliforms	<3/100 ml	must not be consecutive samples
Piped supply of water in distribution system	Fecal coliforms	0/100 ml	In 95 per cent of samples per year
	Total coliforms	0/100 ml	In remaining 5 per cent, which
	Total coliforms	3/100 ml	must not be consecutive samples
Unpiped supply	Fecal coliforms	0/100 ml	Should not occur frequently
	Total coliforms	<10/100 ml	
Bottled water	Fecal coliforms	0/100 ml	Source should be free from fecal contamination
	Total coliforms	0/100 ml	
Emergency supply	Fecal coliforms	0/100 ml	Boiling where this guideline cannot be met
	Total coliforms	0/100 ml	

It must be appreciated that this reliance on measurement of total coliforms and fecal coliforms assumes that they are always present in sewage and absent from environments that are free from sewage. This is almost always true of fecal coliforms, but other types of coliforms are more widely distributed and may lead to erroneous condemnations of safe supplies. Where there are doubts about the coliform results, other tests such as the enumeration of faecal streptococci or anaerobic clostridia may be used as confirmatory tests together with sanitary surveys to discover any possible sources of contamination.

The routine examination for other types of microbial pathogens such as viruses or protozoa presents formidable difficulties. Techniques are available for virus examination, but they require sophisticated laboratory facilities. Instead, reliance is usually placed on the maintenance of a sufficient chlorine residual.

The number of bacteria and other microorganisms is considerably reduced during water treatment. Processes such as coagulation, sedimentation, and filtration cannot be relied upon to achieve a satisfactory reduction, and the final stage in most water treatment plants is disinfection.

The aims of disinfection are twofold: to reduce the bacterial concentration to an acceptable level and to provide residual protection in case of contamination in the distribution system. These goals are generally achieved by dosing the water with chlorine in either the form of chlorine or hypochlorite. The technical problems of handling liquefied chlorine restrict its use to large treatment plants, and there are drawbacks to using chlorine in any form since it reacts with many trace organic substances present in the water and is liable to form carcinogenic products. There are, however, considerable advantages to using chlorine, notably its reaction with ammonia to form chloramines. These are relatively satable and yet bacteriostatic compounds that serve to provide residual protection.

Effective disinfection with chlorine requires low turbidity [<5 nephelometric turbidity units (NTU)] since the bacteria may be inside particles and thus escape contact. Also the pH must be sufficiently low (*i.e.*, <8) to allow the formation of unionized hypochlorous acid. Since chlorination either with chlorine or hypochlorite alters the water's pH, some correction may be required.

Alternatives to disinfection by chlorine are to use ozone or chlorine dioxide or to disinfect physically by using ultraviolet light. Ozone is an attractive alternative since it has a powerful oxidizing action that helps with colour removal and is also more toxic to viruses. It is, however, short-lived and provides no residual protection; neither do chlorine dioxide and ultraviolet light.

Effective disinfection of rural water supplies is often difficult to arrange due to lack of proper supervision. In these circumstances slow-sand filtration is the preferred method of water treatment. It can achieve bacterial reductions of 99.999 per cent, comparable with those normally obtained from disinfection. The process depends upon the buildup of a slime layer around the sand particles. The slime layer develops slowly over a period of weeks and months and requires attention only intermittently when it becomes blocked.

Slow-sand filters are also effective in reduction of turbidity but are not suitable for treating waters of high turbidity due to rapid blockage. They require a much greater area than rapid gravity or pressure filters and are therefore less suitable for large urban treatment plants.

The esthetic criteria for drinking waters are equally as important as the toxic and microbiological criteria since they are concerned with the acceptability of a supply. The guidelines for the esthetic criteria are summarized in Table 1.12. These are more difficult to define precisely since the threshold of tolerance for colour, taste, and so forth varies not only with individuals, but also to some extent with the quality of alternative supplies.

IV. WASTEWATERS

A. Sources and Collection

Wastewaters arise in various ways from human activity. They may be broadly classified as domestic, industrial, agricultural,

TABLE 1.12

WHO Guidelines for Esthetic Quality of Drinking Supplies

Constituent or characteristic	*Guideline value*	*Remarks*
Aluminium	0.2 mg/litre	
Chloride	250 mg/litre	
Chlorophenols and chlorobenzene	—	No guideline value but may affect taste
Colour	15 TCU	
Copper	1 mg/litre	
Detergents	—	No guideline values but may cause foaming or taste
Hardness	500 mg/litre (as $CaCO_3$)	
Hydrogen sulphide	Not detectable by consumers	
Iron	0.3 mg/litre	
Manganese	0.1 mg/litre	
pH	6.5-8.5	
Sodium	200 mg/litre	
Solids, total dissolved	1000 mg/litre	
Sulphate	400 mg/litre	
Taste and odour	Inoffensive to most consumers	
Turbidity	5 NTU	Preferably <1 for effective disinfection
Zinc	5 mg/litre	

urban drainage, and leachates from tips. It is customary to distinguish solid wastes and wastewaters and to treat and dispose of these different types of wastes in different ways, but sludges are intermediate in character. Wastewaters (with the exception of sludges) flow sufficiently easily to be collected by a surface channel and underground pipe system (sewers). Sewer systems convey domestic and industrial wastes (foul sewers) that may be separate from urban drainage (storm sewers), or the two types of sewers may be conveyed in a single system (combined sewers). The design of a sewer system is a difficulty hydraulic problem because of the need to provide self-scouring velocities under varying flows in a system that usually operates under gravity.

Another useful distinction is between point sources (*i.e.*, wastes that can be economically collected together for treatment or disposal and diffuse sources. The latter include agricultural wastes and leachates. In many parts of the world due to severe economic constraints, it is not possible to collect domestic wastes. Also in such communities industries may be in small and widely scattered units. Control of water pollution in such communities relies upon nonsewered devices such as septic tanks and pit latrines.

B. Characteristics

The composition of wastewater needs to be assessed in relation to the following considerations:

1. suitability for discharge to a sewer, whether the wastewater is liable to damage the fabric of the sewer or to cause risk to health for people working in the sewer;
2. suitability for treatment, whether the wastewater is liable to cause difficulties in the treatment process due to high organic strength, nutrient imbalance (*e.g.*, C:N:P not sufficient), toxicity (*e.g.*, heavy metals), foaming, or high suspended soils; and
3. suitability for discharge after treatment, whether the wastewater is liable to cause problems in receiving

waters even after treatment due to the presence of recalcitrant substances or due to high levels of pathogenic organisms.

Domestic wastes vary somewhat in composition around the world depending upon diet, especially water consumption. In Europe the daily production is around 0.06 to 0.08 kg of BOD and 0.8 to 1.0 kg of SS, which with a daily water usage of 180 litres per person results in a sewage of 300 mg/litre of BOD and 400 mg/litre of SS. The corresponding figures in the United States are about 25 per cent less due to higher water usage.

In other parts of the world the composition of sewage is more varied, tending particularly in arid rural areas to be stronger owing to low water consumption and in tropical areas tending to be septic.

The composition of wastes from urban areas also depends upon the pattern of surface water drainage. Where combined sewers are used, very high flows of low-strength wastes can result, especially in tropical regions. Where separate sewers are used, the flow of domestic wastes is much more even but still has a pronounced diurnal pattern, as shown in Fig. 1.8. These fluctuations are important not only in sewer design, but also because they impose varying loads on the wastewater treatment plant.

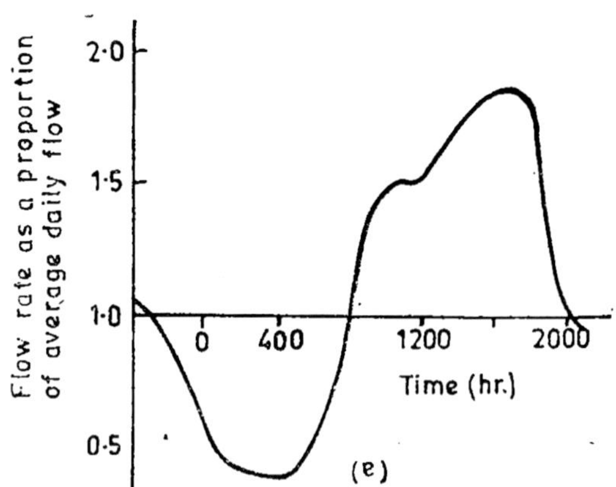

Fig. 1.8: Diurnal variations in the flow of domestic wastes

Apart from organic strength and suspended solids, the most important constituents of sewage are nitrogen, phosphorus, and intestinal bacteria. The nitrogen concentration in raw wastes varies from 60 to 100 mg/litre, of which 20 to 30 mg/litre is ammoniacal and the remainder in the form of organic compounds, mostly urea. The per capita production of phosphorus is about 2 g/day and is much lower than the 9 g/day of nitrogen. The concentration of phosphorus in raw domestic waste is correspondingly less (about 30-40 mg/litre), of which about half is organic and the rest is orthophosphates. The concentration of nitrogen and phosphorus in sewage is more than ample for the nutritional needs of the bacteria that are used in treating wastewaters. The ratio of BOD to nitrogen should be not more than 100:10, and the ratio of BOD to phosphorus should not be more than 100:1. The ratio in domestic waste is usually 100:20:2. The bacterial content of domestic wastes is extremely high, as shown in Table 1.13.

TABLE 1.13

Bacterial Content of Domestic Wastewaters

Organism	*Concentration range*
Total coliforms	10^7—10^9/100 ml
Fecal coliforms	10^7—10^8/100 ml
Fecal streptococci	10^5—10^6/100 ml
Salmonellae	0—10^2/100 ml

The composition of industrial wastes is much more varied than domestic sewage, so much so that it is impossible to characterize it briefly. Table 1.14 shows the composition of only some of the more important types of industrial wastewaters.

The discharge of industrial wastewaters may be more varied than domestic sewage, especially in industries that use batch processes or where factories open for only certain periods, (*e.g.*,

8-hr/day or 5 days/week). Fluctuations in flow may also be accompanied by fluctuations in quality. Despite the variability of industrial wastewaters, it is possible to make a number of generalizations, among them that industrial waste-waters usually contain much fewer bacteria than domestic sewage and that they often contain many more toxic substances.

TABLE 1.14

Composition of Some Common Industrial Wastewaters[1]

Type of wastewater	*Approximate composition*			
	BOD	*SS*	*N*	*P*
Brewery	800—1000	600—800	30—40	8—10
Cannery	500—5000	200—2000	100—500	5—50
Metal processing	100—300[2]	100—500	10—20	2—10
Coke production	500—2000[2]	200—400	100—300	5—20

[1]All concentrations are in milligrams per litre.
[2]Estimated from COD. Due to the toxic nature of waste, BOD is not an appropriate parameter.

The third main source of wastewater is the surface runoff from urban areas, which is often referred to as storm water. The pulluting material in storm water is a reflection of wastes brought into solution or suspension during the water's passage over the surface. The most outstanding characteristic or urban drainage is its extreme variability, as shown in Table 1.15. Part of this variability is due to differences in the nature of the material on the surface and the length of time of contact, but it is due mainly to the first flush phenomenon, as shown in Fig. 1.9.

Sludges are an especially difficult type of wastewater to dispose of. They are intermediate in character between solid and liquid wastes, with a solid content that may vary from 0.5 per cent up to 10 per cent or even 20 per cent. Sludges arise from many industrial processes (*e.g.*, neutralization of acid wastes with lime). Also biological sludges are produced during the primary and secondary sedimentation in wastewater treatment.

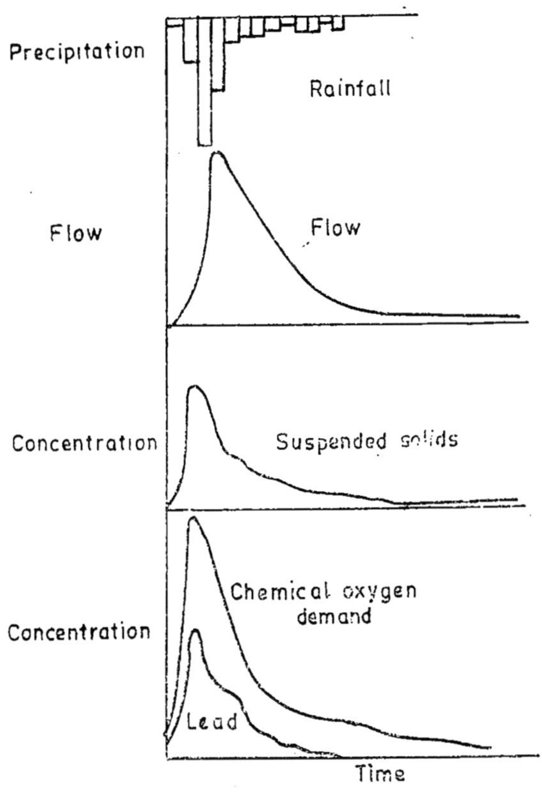

Fig. 1.9: Relationship between flow and quality during a storm

The important characteristics of sludges are the following.

1. *Solids content*: Sludge handling and disposal usually rely upon dewatering the sludge to around 5 per cent to 10 per cent solids to produce a material that can be handled as a solid.

TABLE 1.15

Characteristics of Urban Drainage

Parameter	*Concentration range*
Biochemical oxygen demand	30—500
Suspended solids	20—7000
Total nitrogen	6—70
Total phosphorus	0.1—12
Heavy metals	0.3—10
Total coliforms	10^3—10^8

2. *Organic content*: The amount of degradable organic matter determines the method of disposal and subsequent problems such as odour production.

3. *Pathogen content.* The concentration of bacteria, viruses, helminth eggs, and protozoa cysts can be very high in sludges from wastewater treatment and may require some form of disinfection before disposal.

4. *Toxic content.* Persistent toxins such as heavy metals and chlorinated organics limit the disposal of sludges, especially for agricultural purposes.

BIBLIOGRAPHY

Hynes, H.N.B. (1964). "The Biology of Polluted Waters." Liverpool Univ. Press, Liverpool, England.

James, A. (1984). "An Introduction to Water Quality Modelling." Wiley, London.

Lund, J.W.G. (1971). Eutrophication, In "The Scientific Management of Animal and Plant Communities for Conservation." Symp. Brit. Ecol. Soc. (E. Duffey and A.S. Watt, Eds.). Blackwell, London.

Streeter, H.W., and Phelps, E.B. (1925). A study of pollution and natural purification of the Ohio River, U.S. Public Health Ser., Bull. No. 146.

Vollenweider, R.A. (1968). "Fundamentals of the Eutrophication of Lakes and Flowing Waters." Organization for Economic Corporation and Development, Paris.

World Health Organization (1984). "Guidelines for Drinking Water Quality," 2nd ed. World Health Organization, Geneva.

2

WATER POLLUTION AND TOXICOLOGY

Toxicology is the quantitative study of the harmful effects of chemicals on organisms. The field of toxicology is a broad one and may be broken down into three major divisions, namely, economic, forensic, and environmental toxicology. Economic toxicology is concerned with the deliberate use of toxic chemicals to produce harmful effects on target organisms such as bacteria, parasites, and insects. The obvious applications of economic toxicology are in medicine, agriculture, and forestry management, where it is frequently desirable to eliminate or at least control the numbers of various infectious organisms, parasites, and pests. Forensic toxicology is concerned with the medical and legal aspects of the adverse effects of chemicals on humans. The medical aspects of forensic toxicology concern the diagnosis and treatment of the adverse effects of toxic chemicals, while the legal aspects concern the cause and effect relationships between exposure to a chemical and the effect of this exposure on human health. Environmental toxicology is concerned with the incidental exposure of plants and animals, including humans, to chemicals that are contaminants of the environment, including in particular the water we drink and the water in which aquatic organisms live. Environmental toxicology is therefore directly related to the subject of water pollution because it is environmental toxicological studies that reveal the quantitative relationships between the concentration of chemicals found in the water or in aquatic organisms and the effect of these chemicals on aquatic organisms and on persons who drink the water or consume the organisms.

I. LEGAL CONSIDERATIONS

The quantitative relationships revealed by toxicological studies are ultimately used to set water quality standards. In the United States the principal pieces of federal legislation used to establish water quality standards are the Safe Drinking Water Act (SDWA) of 1974 and the Clean Water Act (CWA) of 1972. SDWA, as the name implies, is concerned strictly with drinking water supplies. The act cells for a set of primary and secondary drinking water standards. The primary standards concern contaminants that the Environmental Protection Agency (EPA) consibers to pose a threat to human health. The secondary drinking water standards apply to contaminants that may adversely affect the odour or appearance of drinking water to such an extent that a substantial number of persons would not drink the water. The CWA is concerned primary with the reduction and control of pollutant discharges into navigable waters, and it is the most comprehensive federal water pollution control programme in existence. Section 303 of the CWA provides that water quality standards be developed for all surface waters. Within the context of the CWA, a water quality standard consists of the following two parts: a "designated use," such as agricultural, recreation, or fish and wildlife for which the body of water is to be protected, and "criteria," which are pollutant concentration limits or some sort of narrative statement required to preserve or achieve the designated use. These water quality standards are established by either federal or authorized state authorities and are used to control the discharge of pollutants into the given body of water. The principal mechanism for controlling pollutant discharges is Section 402 of the CWA, which establishes a National Pollution Discharge Elimination System (NPDES). Under Section 402 the EPA or authorized states can issue permits for pollutant discharges, but can also refuse to issue such permits if the discharge would create conditions that violate the water quality standards developed under Section 303. Although the NPDES applies to point sources, the water quality standards may also form the basis of the best management practices applicable to nonpoint sources uader Section 208 of the CWA. Seotion 208 calls for the development and implementation of area-wide plans for the management of waste treatment to be worked out through cooperative efforts between the EPA and state governments.

Although the CWA is primarily concerned with the reduction and control of pollutant discharges into navigable waterways, both the SDWA and CWA can be used to control groundwater pollution as well. The SDWA, for example, contains an underground injection control (UIC) programme that gives the EPA direct authority over underground waste disposal wells. According to the provisions of the UIC, underground injection is prohibited without an authorized state permit, and to obtain a state permit, the applicant must satisfy the state that drinking water sources will not be endangered by the injection of waste. Section 104 of the CWA requires the EPA to establish and maintain a water quality surveillance system for both navigable waters and groundwaters. Furthermore, Section 402 of the CWA contains a subsection that requires a state to have adequate authority to control the discharge of pollutants into wells as a condition for authorizing a state NPDES programme. Thus, a state with an authorized NPDES programme can control groundwater pollution under the CWA to the extent that the state can control the disposal of pollutants in wells.

The establishment of water quality standards is obviously a critical step in the control of water pollution, and it is in the setting of these standards that information from environmental toxicological studies is utilized. Unfortunately, the establishment of water quality standards has proven to be more difficult and much more time consuming than was envisioned during the 1970s when the SDWA and CWA were passed. For example, under Section 1412 of the SDWA, the EPA promulgated "National Interim Primary Drinking Water Standards" (NIPDWS) for certain organic and inorganic substances in 1975.

Amendments to the NIPDWS were issued in 1976, 1979, and 1980. The NIPDWS establish "maximum contaminant levels" (MCLs), which are the maximum permissable levels of the contaminants in water that may be delivered to a user of a public water system (currently defined as a system serving at least 25 persons). However, as of 1985 the entire NIPDWS list included a total of only ten inorganics, eight organics, turbidity, coliform bacteria and radioactivity (Table 2.1). The EPA is currently in the process of establishing recommended maximum contaminant levels (RMCLs) and revised National Primary Drinking Water Regulations (NPDWRs) for volatile synthetic organic chemicals, synthetic organic chemicals, inorganic chemicals, microbiological contaminants, radionuclides, and disinfectant by-products including trihalomethanes. RMCLs were proposed for nine volatile organic chemicals on June 12, 1984 (Table 2.2). These RMCLs are the levels at which, in the EPA administrator's judgement, no known or anticipated adverse effects on the health of persons occur and which allow an adequate margin of safety. The RMCLs are nonenforceable health goals, but the MCLs, which are enforceable, must be set as close to the RMCLs as is feasible.

On October 5, 1983, the EPA gave advanced notice of proposed rule making (ANPRM) concerning synthetic organic chemicals, inorganic chemicals, microbiological contaminants, and radionuclides. The substances on the proposed list are given in Tuble 2.3, and the schedule of past and anticipated developments in the establishment of MCLs under the SDWA is indicated in Table 2.4. Perhaps the most noteworthy feature of Table 2.4 is that it has required the EPA over 10 years since the passage of the SDWA in 1974 to begin to set MCLs for so many chemicals. Similarly, 5 years after the original version of the CWA was written in 1972, the EPA had failed to draw up any regulations. As a result, in 1977 Congress passed revisions to the CWA that ordered the EPA to regulate 65 specific pollutants. Water quality criteria for 64 of those 65 pollutants were published in the Federal Register of November 28, 1980, and for dioxin (the 65th pollutant) in the Federal Register of February 15, 1984.

TABLE 2.1

National Interim Primary Drinking Water Standards

Contaminant	*Maximum contaminant levels as of 1985*
Inorganics	
Arsenic	50 ppb
Barium	1.0 ppm
Cadmium	10 ppb
Chromium	50 ppb
Fluoride	1.4-2.4 ppm (temp. dependent)
Lead	50 ppb
Mercury	2 ppb
Nitrate (as N)	10 ppm
Selenium	10 ppb
Silver	50 ppb
Organics	
Chlorinated hydrocarbons	
Endrin	0.2 ppb
Lindane	4 ppb
Methoxychlor	100 ppb
Toxaphene	5 ppb
Total trihalomethanes	100 ppb
Chlorophenoxys	
2, 4-D	100 ppb
2, 4, 5-TP silvex	10 ppb
Microbiological	
Coliforms	1/100 ml monthly average 4/100 ml single sample
Turbidity	1 T.U.
Radioactivity	
Gross α	15 pCi/litre
Radium 226+228	5 pCi/litre
β+photon	4 mrem/yr

TABLE 2.2

Recommended Maximum Contaminant Levels (RMCLs) Proposed for Volatile Synthetic Organic Chemicals on June 12, 1984

Chemical	*Proposed RMCL (ppm)*
Trichloroethylene	0
Tetrachloroethylene	0
Carbon tetrachloride	0
1, 1, 1-Trichloroethane	0.2
1, 2-Dichloroethane	0
Vinyl chloride	0
1, 1-Dichloroethylene	0
P-Dichlorobenzene	0.75
Benzene	0

Why does the EPA require so much time to set water quality standards? The answer to this question is undoubtedly in part political, but the difficulty in obtaining and interpreting relevant toxicological information has also been a major problem. The nature of the problem was highlighted in a statement by EPA administrator William Ruckelshaus, who noted,

> We assumed [10 or 15 years ago] we knew what the bad pollutants are, at what levels they cause adverse environmental or public health threats. We [assumed] we knew how to drive them down to a no-effect level, how to measure them, how to control them at a reasonable cost. All you needed was an enforcement presence that was sufficiently strong. [Except for the need for enforcement,] all those assumptions were wrong. [Sun, M. (1985). Legislative paralysis on the environment. Science 227, 496-497.]

TABLE 2.3

Contaminants for Which Advanced Notice of Proposed Rule Making was given on October 5, 1983

	Inorganics	
Aluminium	Cyanide	Sulphate
Antimony	Molybdenum	Thallium
Asbestos	Nickel	Vanadium
Beryllium	Sodium	Zinc
Copper		
	Synthetic Organic Chemicals	
Acrylamide	1, 2-Dichloropropane	PCBs
Adipates	Dinoseb	Pentachlorophenol
Alachlor	Dioxin	Phthalates
Aldicarb	Diquat	Pichloram
Atrazine	Endothall	Simazine
Carbofuran	Epichlorohydrin	Toluene
Chlordane	Ethylene dibromide	1, 1, 2-Trichloro-ethane
Dalapon	Glyphosate	
Dibromochloropropane (DBCP)	Hexachlorocyclopenta-diene	Vydate Xylene
Dibromomethane	PAHs	
	Microbiological Contaminants	
Total Coliforms		
	Radionuclides	
Gross α	Radon	
β+photon Radium 226+228	Uranium	

TABLE 2.4

Actual and Proposed Regulatory Development Schedule for Drinking Water Regulations under the SDWA

Category of Contaminants	*ANPRM*	*Proposed RMCL*	*Promulgate RMCL propose MCL*	*Promulgate MCL*
Volatile organics	3/4/82	6/12/84	4/85	4/86
Soluble organics, inorganics, microbiology	10/5/83	5/85	2/86	12/86
Radionuclides	10/5/83	12/85	—	—
Disinfectant by-products	1986	—	—	—

Ruckelshaus' statement emphasizes the critical role environmental toxicology plays in setting water quality standards and that the role has changed significantly in recent years. Using toxicological information to set water quality standards is now perceived as a far more complex problem than was thought to be the case in the early 1970s when the CWA and SDWA were passed. What sort of toxicological information is needed to determine whether water is suitable for an intended use, and how is that information used to set water quality criteria? [See *Pollution, Environmental*]

II. TOXICITY TESTS

In a general sense toxicity tests may be assigned to one of two categories. The first category involves studies of the general overall effects of a compound on an organism. Such overall effects may be described by one or more of the following terms:

Acute. involves a stimulus severe enough to bring about a response speedily, usually within 96 hr for fish.

Subacute. involves a stimulus less severe than an acute stimulus, producing a response in a longer time; may become chronic.

Chronic. involves a lingering or continuous stimulus; often signifying periods of about one-tenth of the life span or more.

Lethal. causes death by direct action.

Sublethal. insufficient to cause death.

Cumulative. brought about, or increased in strength, by successive additions.

These general effects can be broadly classified under the subcategories of acute toxicity, which is usually lethal, and chronic toxicity, which may be lethal or sublethal. From the standpoint of testing, Loomis (1974) defines three types of tests associated with general toxicological effects Acute tests involve a single exposure of the test organism to the toxicant and are usually intended to reveal the symptoms associated with exposure to the toxicant and to determine the degree of lethality of the toxicant. Prolonged toxicity tests involve repeated exposures of the experimental organism to the toxicant and are usually conducted over a period of several months. Prolonged toxicity tests are usually intended to allow a thorough evaluation and characterization of all effects of the toxicant on the experimental organism. Chronic toxicity tests are usually conducted for a period of a year or more and are performed for one of two purposes. First, they may be intended to determine the degree of exposure that produces no demonstrable effect on the test organism even after an extended period of exposure. Second, they may be designed to determine the carcinogenic protential of the toxicant.

The second category of toxicity tests involves experiments designed to evaluate in detail specific kinds of toxicity. These specific kinds of tests may be designed to study the interaction between two or more toxic substances and to study the tendency of a toxicant to cause abnormal development of the fetus (teratogenic tests), to affect the reproductive capacity of an organism, to cause mutations (mutagenic tests), to produce tumors (tumorigenic tests), to cause cancer (carcinogenic tests), to affect the photosynthetic rates of plants, and so forth. These specific kinds of tests are necessary because many of the important effects of toxicants on organisms, particularly sublethal effects, simply do not become apparent in standard tests to evaluate overall effects.

A. Sublethal Effects

Concentrations of a toxicant that are insufficient to kill organisms outright may nevertheless have a devastating effect on the population of the same organism over a sufficient period of time as a result of sublethal effects. Although these sublethal effects may be highly variable, in general they may be classified as modifications or interferences with reproduction, development or growth, and behaviour. The aim of setting water quality standards is to eliminate or at least minimize even the subtle sublethal effects of pollutant substances. Unfortunately, the task of determining the level of stress that causes no significant sublethal effects has proven much more difficult to determine than was once imagined, in part because of the large number of potentially toxic substances and because of the large number of possible sublethal effects. In some cases simply designing an experiment to study certain types of sublethal effects is no trivial matter. A few examples will illustrate the variety and complexity of sublethal effects.

1. Reproduction

Obviously any pollutant that interferes with or blocks the reproduction of a species may completely eliminate that species without having any apparent effect on the adult members of the population. For example, DDE, which is a metabolite of DDT, was implicated as the cause of the reproductive failure of certain fish-eating birds, such as the pelican, during the 1960s and early 1970s. Certain toxicological tests designed to examine the specific effects of DDE on the reproduction of various birds showed convincingly that high levels of DDE in the feed of female birds lowered the level of estrogen in the birds and caused them to lay thin-shelled eggs. Table 2.5 illustrates the results of one such study.

TABLE 2.5

Reproduction and Eggshell Thickness Data from Penned Mallard Ducks Maintained for Two Seasons on Feed Containing DDE[1]

DDE added to feed (ppm)	*No. of 14-day ducklings per hen (% of control results)*		*Shell thickness (% of control results)*	
	Year 1	*Year 2*	*Year 1*	*Year 2*
10	51	24	92	89
40	35	20	87	86

[1]From Health, R.G., Spann, J.W., and Kreitzer, J.F. (1969), Marked DDE impairment of mallard reproduction in controlled studies. *Nature* (*London*) 224, 47-48.

In this particular experiment, which lasted over a period of two years, test ducks began receiving feed containing either 10 or 40 ppm DDE several weeks before the onset of the first laying season. A control group of ducks received feed containing no added DDE. The ducks fed 10 or 40 ppm DDE produced 49-80 per cent fewer 14-day ducklings per hen than the control group and produced eggshells that were about 10 per cent thinner than the control eggshells.

That fish-eating birds might have encountered food containing as high as 10 ppm DDE during the 1960s is confirmed by the data in Table 2.6. In 1969 Southern California anchovies, which are a principal food source of the Southern California brown pelican, were found to contain an average of 3.24 ppm DDE on a fresh weight basis. On a dry weight basis this concentration is equivalent to a figure of about 9.7 ppm and is therefore virtually identical to the concentration of 10 ppm dry weight, which Table 2.5 indicates reduced the reproduction of mallard ducks by 50 to 80 per cent. The sharp drop in anchovy

DDE content between 1969 and 1970 coincides with the time that the Montrose Chemical Company (located in Los Angeles County and the sole manufacturer of DDT in the United States) began disposing of its liquid wastes in a sanitary landfill (April 1970) rather than discharging the wastes into the municipal sewer system. The DDE content of the anchovies then remained constant for about 3 yr, but dropped sharply between 1972 and 1973 when the EPA ban on DDT use in the United States went into effect (January 1973). The EPA ban resulted in no small part from the examination of toxicological results such as those shown in Table 2.5, which convincingly demonstrated the adverse effects of DDE on the reproduction of aquatic birds.

TABLE 2.6

Geometric Mean Residues of DDE in Anchovies and Brown Pelican Eggs Off the Southern California Coast[1]

Year	*Anchovy whole bodies (ppm, fresh weight)*	*Intact brown pelican eggs (ppm, lipid weight)*
1969	3.24	853
1970	0.84	—
1971	0.87	—
1972	0.74	221
1973	0.18	175
1974	0.72	97
1975[2]	—	113

[1]Data from Anderson, D.W., Jehl, J.R., Risebrough, R.W., Woods, L.A., Deweese, L.R., and Edgecomb, N.G. (1975). Brown pelicans: improved reproduction off the southern California coast. *Science* 190, 806-808. Anderson, D.W., Jurek, R.M., and Keith, J.O. (1977). The status of brown pelicans at Anacapa Island in 1975. *Cal. Fish Game* 63, 4-10.

[2]Data for 1975 for Anacapa and Santa Cruz island only.

The correlation between DDT use and the reproductive success of the Southern California brown pelican is illustrated in Table 2.7. In 1969 and 1970 only nine fledglings were produced from a total of 1852 nesting attempts, and the thickness of intact eggshells averaged about 0.40 mm. In 1971 and 1972 following the diversion of Montrose's liquid waste to a sanitary landfill, reproduction improved to 249 fledglings out of 1161

TABLE 2.7

Recent History of Brown Pelicans Breeding Off the Coast of Southern California and North Western California[1]

Year	*No. of nests built*	*No. of young total[2]*	*Fledged per nest*	*Intact egg shell thickness (mm)*	*Anchovy abundance*
1969	1125	4	0.004	0.402	140
1970	727	5	0.007	0.393	70
1971	650	42	0.065	0.460	80
1972	511	207	0.405	0.438	195
1973	597	134	0.225	0.510	275
1974	1286	1185	0.922	0.482	355
1975[3]	292	256	0.88	0.51	—

[1]Data from Anderson *et al.*, (1975, 1977). See footnote a of Table 2.6.
[2]Productivity totals include Anacapa Island, Santa Cruz Islands, and Isle Coronada Norte.
[3]Data for 1975 for Anacapa and Santa Cruz islands only.

nesting attempts, and eggshell thickness increased to about 0.45 mm. In 1973-1975 following the ban on DDT use in the United States, reproductive success improved further to 1575 fledglings out of 2175 nesting attempts, and intact eggshell thickness averaged 0.50 mm. The most recent data have shown almost no evidence of crushed eggs due to thin eggshells, but there is now concern that commercial overfishing of anchovies may be adversely affecting the recovery of the Southern California brown pelican due to the reduction in its food supply. Because of this additional stress, it seems likely that this bird would rapidly have become extinct if the disastrous reproductive failures of 1969-1970 had continued for many more years. It is noteworthy, however, that the levels of DDE in the adult pelicans apparently had no effect on the birds' ability to carry out activities other than reproduction in a normal fashion. Fortunately, careful toxicological studies of the sublethal effects of DDE on bird reproduction had been carried out and were presented to the EPA at the time of the 1972 DDT hearings. These studies did much to influence the EPA decision to ban DDT use in the United States.

2. *Development and Growth*

Interference with the normal physiological processes of an organism may adversely affect its ability to grow and, develop without directly killing the organism. Nevertheless, because of the pressure of competition and predation, a population of organisms whose development or growth has been retarded may be rapidly eliminated from an ecosystem. For example, studies have shown that exposure of certain species of marine phytoplankton to mercury concentrations as low as 0.5 parts per billion (ppb) can cause more than a 50 per cent reduction in photosynthetic rates. Figure 2.1. illustrates some of the experimental results obtained with the diatom Nitzschia delicatissima. If the presence of a toxin in the water significantly reduces the photosynthetic rate of a particular species, it is possible that the species will be grazed to extinction by herbivores without being directly killed by the toxin. Similarly, phytoplankton division rates and/or photosynthetic rates have been found to be reduced at polychlorinated biphenyl (PCB) concentrations as low as 0.1 to 10 ppb. The effect of PCB on phytoplankton differs markedly between species, some species being much more sensitive than others. As a result, the species composition of a phytoplankton community may be greatly altered due to the presence of PCB in the water at concentrations as low as 0.1 ppb. Studies in a natural estuarine marsh showed that exposure to as little as 1.0

ppb PCB reduced phytoplankton production and caused a shift in the size structure of the phytoplankton community toward smaller cells. This shift in the mean cell size of the phytoplankton could effectively increase the number of trophic levels between the primary producers and commercially useful fish and/or could divert more production toward various gelatinous predators such as jellyfish. Thus, the sublethal effects of a toxin on organisms at the base of the food chain could have a profound effect on the structure of the entire food chain and could significantly affect the productivity of the system with respect to commercially useful organisms.

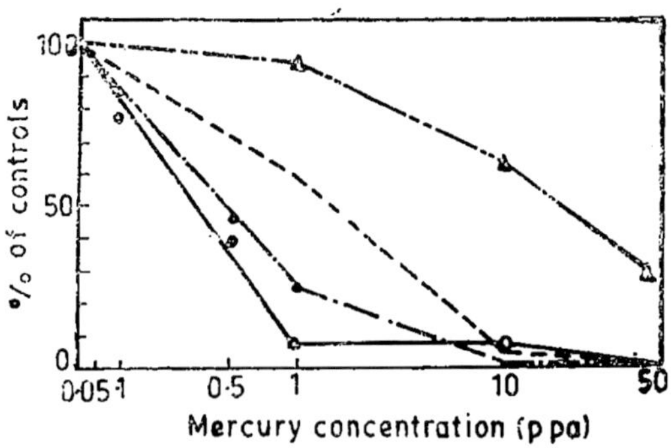

Fig. 2.1: Photosynthesis by the marine diatom Nitzschia delicatissima after 24 hr of exposure to the following mercurials: solid triangles, diphenyl mercury; open triangles, phenylmercuric acetate; solid circles, methylmercury dicyandiamide; open circles, MEMMI (a compound containing methylmercury). [Reprinted with permission from Harriss, R. C., White, D. B., and MacFarlane, R. B. (1970). Mercury compounds reduce photosynthesis by plankton. Science 170, 736-737. © 1970 AAAS.]

One of the best documented effects of a stress on the development and growth of aquatic organisms is that of temperature. All organisms grow most efficiently over a limited temperature range. Temperatures above or below the optimal range lead to reduced growth rates or to reduced growth efficiency. Figure 2.2 illustrates in a qualitative way how temperature can influence the growth rate and efficiency of an organism. In the figure A_c is the amount of food consumed by the organism, and A_g is the amount of that food converted into biomass (growth). A_w is the amount of food that is excreted (waste), while A_a, A_d, and A_S represent the amounts of food that are respired to support the organism's activity (A_a), the degradation and metabolism of consumed food (A_d), and standard or basal metabolism (A_S). It is clear from Figure 2.2 that there is some temperature at which growth (A_g) is a maximum, because the rate of food consumption (A_c) peaks at a certain temperature, whereas the basal respiration rate (A_s) continues to rise exponentially. This dependence of A_c and A_s on temperature is characteristic of the behaviour that might be expected from a poikilothermic (coldblooded) organism. The

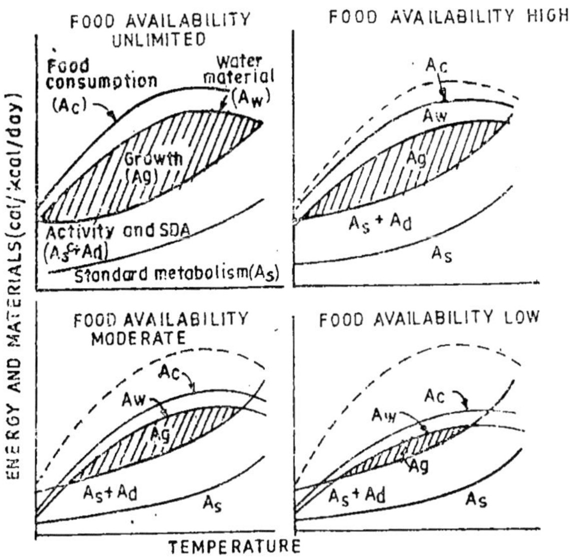

Fig. 2.2: Effects of temperature and food availability on food consumption (A_c) and the partitioning of consumed food for various purposes by a poikilothermic animal. The curves are qualitative and purely theoretical. [Reprinted with permission from Warren, C. E. (1971), "Biology and Water Pollution." Saunders, Philadelphia, © 1971 W. B. Saunders Co.]

amount of food converted to biomass is positively correlated with temperature only as long as the difference between A_c and A_w increases more rapidly than A_s.

Figure 2.2 illustrates one of the difficult problems encountered in converting toxicologic information into water quality standards. In all cases shown, it is clear that a small deviation in temperature from the optimal value will result in some decrease in growth, but this effect by itself will not be lethal. However, the effect of a change in temperature on growth is obviously dependent on the availability of food to the organism. A change of temperature that would cause only a modest reduction in growth when the food supply is unlimited could be lethal (result in negative growth) when food availability is low. Furthermore, although a modest reduction in growth might not by itself be lethal, in a highly competitive system, even a small reduction in growth could lead to the elimination of a species through the combined effects of competition and predation. Similarly, the ecological impact of a given temperature stress might be far greater in a system in which the organisms were already subjected to other forms of stress (for example, stress caused by the presence of toxic chemicals in the water) than in a system in which no other stresses existed. Thus, translating apparently straightforward toxicological information into water quality standards can be quite difficult.

3. Behaviour

Modification in behaviour patterns may involve a wide variety of effects, such as alterations in migratory behaviour, learning ability, feeding behaviour, ability to avoid predators, and so forth. A few examples will illustrate some of the kinds of problems that pollutants may cause as a result of behavioural modification.

Salmon migrate from the ocean back to the freshwater stream where they were born in order to spawn. Although the mechanism that these fish use to find their way from the open ocean back to the general vicinity of their home stream is not well understood, experiments have clearly shown that salmon use their sense of smell to guide their migration upstream once they have entered the freshwater system. The chemical substances that guide this migration have been shown to include certain volatile organic compounds. Unfortunately, a number of streams once characterized by large salmon runs have become seriously polluted with industrial wastes, particularly in the northeastern United States. In many of these streams salmon runs are now no longer observed or involve insignificant numbers of fish. There is good reason to believe that the presence of industrial wastes, including a variety of organic substances, in these streams has so altered the characteristic smell of the water that the fish are no longer able to identify their home stream. Thus, the presence of chemicals in the water at concentrations well below lethal levels may completely disrupt the migratory patterns of the salmon and, through this behavioural modification, effectively eliminate the species.

Seasonal changes in water temperature can lead to a variety of natural behavioural responses in aquatic organisms. These responses include the onset of reproductive cycles and migratory behaviour. The discharge of heated wastewater from large electric power plants can completely disrupt these temperature cues and therefore seriously interfere with natural behaviour patterns. For example, menhaden found in the coastal waters of New Jersey during the summer months normally migrate to warmer waters off the North Carolina coast during the winter. At Barnegat Bay, New Jersey, the Jersey Central Power and Light Company operates a nuclear power plant that discharges cooling water at a temperature about 14°C above ambient. During the winter, large numbers of menhaden are often found in the waters near the Oyster Creek outfall, either because they are naturally attracted to the warm water in the winter and/or because, finding themselves in the warmer waters near the outfall, they do not receive the proper temperature cue to migrate south. In this case the migratory behaviour of the menhaden is completely altered by the presence of the heated effluent water, which by itself has no adverse physiological effect on the fish. In fact, during the winter months the temperature of the effluent water is undoubtedly much closer to the optimum temperature for the menhaden than is the temperature of the surrounding water. Unfortunately, the water temperature near the outfall can drop quickly back to ambient levels if the power plant is shut down. Such a shutdown occurred on January 28, 1972, and as a result 100,000-200,000 menhaden died of thermal shock. In a similar incident in Long Island Sound approximately 10,000 bluefish that had been attracted to the heated plume from the Northport power plant died from thermal shock when winds and tidal currents caused the plume to suddenly shift position on January 17, 1972. Similar incidents can be anticipated in the future wherever the heated wastewater from power plants attracts aquatic organisms during the colder months of the year.

Pollutant effects on the learning ability of organisms can be a particularly subtle form of sublethal stress. An experiment performed on conditioned goldfish provides a good example of how learning ability can be impaird by low levels of pollutants, and at the same time illustrates the dramatic difference between lethal and sublethal concentrations of pollutants. The goldfish were conditioned by teaching them to respond when a light was turned on in one side or the other of the aquarium in which they were kept. If the goldfish remained in the half of the aquarium with the lighted bulb for more than 3 sec, they received an electrical shock. If they swam to the darkened side of the aquarium, they received no shock. The goldfish were trained to respond to the light until the percentage of goldfish responding correctly varied by no more than 5 per cent for 3 consecutive days. Once trained in this way, the goldfish responded correctly about 69 per cent of the time.

After this training period, the goldfish were put into tanks containing various concentrations of either arsenic, lead, mercury, or selenium for a total of 48 hr. After 24 hr and again after 48 hr, the goldfish were retested to check their ability to respond correctly to the light. After the 48-hr-exposure period, the goldfish were returned to clean water and were retested every 24 hr until their percentage of correct responses varied by no more than 5 per cent for 2 consecutive days. If this final score was found to be significantly lower than the score the fish had made prior to exposure to the metal, then the metal was concluded to have adversely affected the fishes' learning ability.

Table 2.8 lists the lowest concentration of each metal that significantly impaired the learning ability of the goldfish. For comparative purposes, the concentration of each metal that proved to be lethal to the goldfish is listed in the second column of the table, and the third column gives the sublethal concentration as a percentage of the lethal concentration. A metal concentration was considered to be lethal if 50 per cent of the goldfish died during a 48-hr exposure to the given concentration. The most dramatic aspect of the data in Table 2.8 is the large difference between the lethal and sublethal concentration. The sublethal concentrations are 50 to 1700 times smaller than the corresponding lethal concentrations. Similar experiments with other organisms and other pollutants have shown that sublethal stresses can often be detected at concentrations that are several orders of magnitude smaller than lethal concentrations. The ratio of concentrations that are lethal and sublethal has proven to be an important parameter in establishing water pollution standards because in many cases there is far more information available on lethal toxicity than on sublethal toxicity. Although water quality standards are intended to be

TABLE 2.8

Lowest Metal Concentration Found to Impair the Learning Ability of Conditioned Goldfish[1]

Metal	*Sublethal concentration (ppm)*	*Lethal concentration (ppm)*[2]	*Sublethal/lethal (per cent)*
Arsenic	0.1	32	0.31
Lead	0.07	115	0.06
Mercury	0.003	81	0.37
Selenium	0.25	12	2.08

[1]From Weir, P. A., and Hine, C. H. (1970). Effects of various metals on behaviour of conditioned goldfish. Arch. Environ. Health 10, 45-51.
[2]48-hr TLm.

criteria that will provide reasonable protection against sublethal stresses, in many cases there is simply not enough information on sublethal toxicity effects to warrant setting a standard. In such cases a lethal toxicity standard is established based on an adequate amount of lethal toxicity tests, and an average lethal/sublethal toxicity ratio is then calculated based on a smaller amount of information. Finally, a sublethal toxicity standard is established by dividing the lethal toxicity standard by the lethal/sublethal ratio. The rationale for this procedure is that for a given pollutant the lethal/sublethal ratio is likely to be far more constant between species than is the sublethal stress level itself.

III. ACUTE TOXICITY DETERMINATION

There is a large amount of information available on the lethal effects of pollutants on aquatic organisms. This fact arises in part because experiments designed to test the lethality of a substance are relatively easy to perform. On the other hand, there are a large number of potentially serious sublethal effects, and experiments designed to examine these effects are often more difficult to design and more time consuming to perform than lethal toxicity experiments.

In aquatic toxicology lethal toxicity tests are often an important component of acute toxicity testing. In designing these experiments, the following two factors must always be taken into consideration: the natural variability among the members of a species in tolerance to the given stress and the relationship between the duration of exposure and the effect of the stress. We consider the problem of natural variability first.

A. Median Survival Times

Because of the natural variability among members of a given species, any sample of that population can be expected to contain individuals that differ in their tolerance to a given stress. How then can the response of the population to the stress be quantified? Should one use the response of the most sensitive individual or the least sensitive individual, or should one use the mean or median response? Present water quality standards are based on the median response of a sample of organisms from the population. Whether one considers this policy conservative or liberal depends to a certain extent on one's frame of mind. The industrialist might argue that the time required to kill all the organisms should be termed the survival time, and that the median survival time is unduly conservative because half the organisms are still alive at the end of the experiment. The environmentalist might argue that the longest time interval during which all the organisms survive should be termed the survival time, and that the median survival time is unduly liberal because half the organisms are dead at the end of the experiment. However, one should keep in mind that median survival times do not themselves become the basis for long-term water quality standards. Ultimately, it is chronic toxicity studies that determine the long-term water quality standards. Furthermore, median survival times are far more reproducible from one sample of a population to the next than are either of the extreme survival times which the industrialist or environmentalist might prefer.

Figure 2.3 illustrates how median survival times are calculated. The data are taken from a study of the toxicity of low oxygen levels to juvenile brook trout. The trout were conditioned for a period of time in water containing 10.5 mg/litre dissolved oxygen and were then placed in experimental aquaria having lower concentrations of dissolved oxygen. The survival of the trout was then monitored over a period of 2000 min. As each fish died, it was removed from its aquarium and its time of death recorded. A total of ten fish were studied at each oxygen level. The results were plotted on semilog paper, and the data fit with straight lines. At the two highest oxygen levels, over half the fish survived for the entire 2000 min, and no effort was therefore made to estimate median survival times at these oxygen concentrations. However, at the eight lowest oxygen levels, over half the fish died during the study period, and estimates of median survival time were made from the intercept of the regression lines fit to the data, with the horizontal line at 50 per cent mortality. This graphical technique of determining median survival times is common in acute toxicity studies.

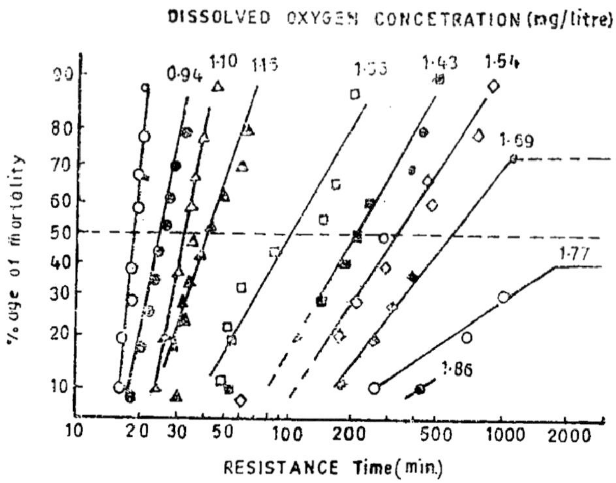

Fig. 2.3: Percentage mortality versus resistance time for juvenile brook trout exposed to the indicated concentrations of dissolved oxygen after being conditioned 10.5 mg/litre oxygen. [Reprinted with permission from Shepard, M.P. (1955). Resistance and tolerance of young speckied trout (Salvelinus fortinalis) to oxygen lack, with special reference to low oxygen acclimation. J. Fish. Res. Bd. Canada 12, 347-446. @ 1955 Fisheries Research Board of Canada.]

B. Incipient Lethal Levels

Given the relationship between median survival time and pollutant concentration, it is possible to calculate lethal concentrations, but now explicit recognition must be taken of the relationship between survival time and degree of exposure. There are basically two approaches. In one approach a graph is made of median survival time versus toxicant concentration, and the graph is extrapolated to time infinity. The hypothetical concentration of the toxicant corresponding to an infinite median survival time is called the incipient lethal level. Fig. 2.4 illustrates how incipient lethal levels are calculated. The data are taken from a study on the toxicity of zinc to rainbow trout. Median survival times were calculated as a function of zinc concentration at three different levels of water hardness. Note that both axes are plotted on a logarithmic scale. At water hardness levels of 50 and 320 mg/litre, the data show distinct curvature, and a function of the form $(C-C_i)^n(T-T_t)=K$ was fit to the data. In this equation C is the concentration of zinc corresponding to median survival time T. As T approaches infinity, C approaches C_i, and C_i is therefore interpreted as the incipient lethal level. The values of C_i estimated from this curve fitting were 3.5 and 1.5 mg/litre at water hardness levels of 320 and 50 mg/litre, respectively. At a water hardness of 12 mg/litre, it was impossible to estimate an incipient lethal level because the data are linear over the range of concentrations studied. All that can be said is that the incipient lethal level at 12 mg/litre hardness is less than 0.5 mg/litre zinc.

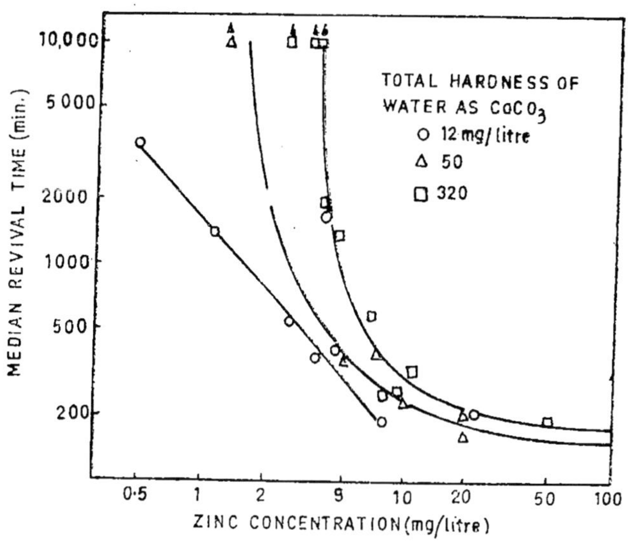

Fig. 2.4: Median survival time of rainbow trout as a function of zinc concentration at three levels of water hardness. [Reprinted with permission from Lloyd, R. (1960). The toxicity of zinc sulphate to rainbow trout. Ann. Appl. Biol. 48, 84-49. @ 1960 Biochemical Society, London.]

It should be obvious that the incipient lethal level is in a sense a hypothetical concept because no organism survives for an infinite time, regardless of the level of stress. However, the incipient lethal level does provide a standard measure of the concentration that produces no acute effects on survival, and it is apparent from Fig 2.4 that at water hardness levels of 50 and 320 mg/litre the concentration of zinc corresponding to a median survival time in excess of about 2000 min is very insensitive to survival time. In other words, at a water hardness of 320 mg/litre, the zinc concentration corresponding to a median survival time of 2000 min and the zinc concentration corresponding to a hypothetical median survival time of infinity are nearly the same. This latter observation leads to the second method for estimating lethal concentrations.

C. Median Tolerance Levels

Rather than extrapolating median survival times to infinity, the toxicant concentration corresponding to a standard survival time is determined. In most cases this survival time is either 48 or 96 hr (2880 or 5760 min). The toxicant concentration corresponding to this median survival time is called the median tolerance limit or TLm. In some cases the abbreviations TL50 (tolerance limit corresponding to 50 per cent survival) or LC50 (lethal concentration corresponding to 50 per cent mortality are used to designate the same parameter. In a more general sense the abbreviation EC50 is often used to designate the concentration that produces a specified effect on 50 per cent of the organisms after a certain period of exposure. In all cases the symbol should be preceded by a designation of the associated period of time. Thus, a 96-hr TLm would be the concentration of a toxicant associated with 50 per cent survival after 96 hr.

Fig. 2.5 illustrates the calculations of a 96-hr TLm. The data are taken from a study of the toxicity of copper to bluegill fish. The fish were divided into two groups and exposed to various concentrations of either $CuSO_4$ or $CuCl_2$ for a period of 96 hr. Percentage survival was then plotted against time. The two data sets were very similar, and a smooth curve was therefore drawn through the combined results. The intersection of this curve with the horizontal line at 50 per cent survival corresponds to the TLm, which in this case was 0.74 mg/litre.

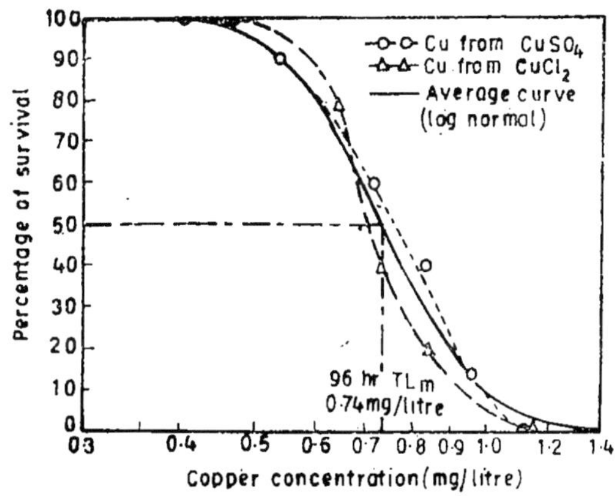

Fig. 2.5: Percentage survival of bluegill fish after 96 hr versus copper concentration. Note that the abscissa has a logarithmic scale. The smooth curve is a log-normal curve adjusted to give the best fit to the combined $CuSO_4$ and $CuCl_2$ data. The intersection of this curve with the horizontal line at 50 per cent survival gives the median tolerance limit (TLm) after 96 hr, 0.74 mg/litre. [Reprinted with permission from Trama, F.B. (1954). The acute toxicity of copper to the common bluegill (Lepomis macrochirus Rafinesque). Acad. Nat. Sci. Philadelphia, Not. Nat. No. 257. (@) Academy of Natural Sciences of Philadeldhia.]

From a scientific standpoint both incipient lethal levels and median tolerance limits are arbitrary measures of lethal concentrations, and there is often little reason to prefer one over the other when setting water quality standards. This fact is apparent in the case of Fig. 2.4, where it is obvious that 96-hr TLm's would be virtually identical to incipient lethal levels at water hardness levels of 50 and 320 mg/litre. At a water hardness of 12 mg/litre, neither an incipient lethal level nor a 96-hr TLm can be estimated because the data are linear and in no case did half the fish survive for 96 hr. From a practical standpoint, however, TLm's are easier to determine because a detailed time series study (*e.g.*, Fig. 2.3) is not required. In determining a TLm it is only necessary to record survival at a single time, usually at either 48 or 96 hr. Calculation of incipient lethal levels does require careful time course studies and, in addition, requires extrapolation of the results to time infinity, a procedure that in some cases may not be straightforward. Perhaps in part because of these considerations, present EPA water quality criteria are written in terms of TLm's rather than incipient lethal levels.

D. EPA Water Quality Guidelines

The rationale and procedures for determining water quality guidelines in the United States are described in the 1980 EPA guidelines.[1] In establishing these guidelines, the EPA has not attempted to suggest standards that would make the water safe for every conceivable organism. The rationale behind the guidelines is summarized as follows:

> Aquatic communities can tolerate some stress and occasional adverse effects on a few species, and so total protection of all the species all of the time is not necessary. Rather, the Guidelines attempt to provide a reasonable and adequate amount of protection with only a small possibility of considerable overprotection or underprotection. Within these constraints, it seems appropriate to err on the side of overprotection. [EPA (1980a),[1] p. 79342.]

Within the context of this statement, it is now useful to examine the minimum data base that the EPA feels is needed for purposes of establishing acutely toxic concentrations. The rationale behind establishing a minimum data base is that,

> Results of acute and chronic toxicity tests with a reasonable number and variety of aquatic animals are necessary so that data available for tested species can be considered a useful

indication of the sensitivities of the numerous untested species.

The minimum data base differs for freshwater and marine organisms and is stipulated as follows. The criterion in freshwater must be based on acute tests with freshwater animals in at least eight different families including at least one representative from each of the following six categories:

1. salmonid fish,
2. nonsalmonid fish,
3. planktonic crustaceans,
4. benthic crustaceans,
5. benthic insects, and
6. benthic detritivores.

For marine organisms, the criterion must be based on acute tests with saltwater animals in at least eight different families subject to the following four constraints:

1. at least two different fish families are included,
2. at least five different invertebrate families are included,
3. either the Mysidae or Penaeidae family or both are included, and
4. at least one of the invertebrate families is in a phylum other than Arthropoda.

If studies have been carried out on an appropriate number and variety of organisms, it may be possible to establish what the EPA refers to as the "final acute value" of the given toxicant, but additional conditions must still be met. The duration of exposure to the toxicant must have been appropriate to the life stage of the species in every case. The following four conditions must be met:

1. For first-instar (less than 24-hr-old) daphnids and other cladocerans, and second-or third-instar midge larvae, EC50 values for immobilization and TLm values must be based on 48-hr experiments.
2. For embryos and larvae of barnacles, bivalve molluscs (clams, mussels, oysters, and scallops), sea urchins, lobsters, crabs, shrimps, and abalones, EC50 values for incomplete shell development and TLm values must be based on 48-to 96-hr experiments.
3. For oysters, EC50 values for decreased shell deposition must be based on 96-hr experiments.
4. For aquatic animals except cladocerans, midges, and animals whose behaviour or physiology allows them to avoid exposure to a toxicant or for whom the acute adverse effect of the exposure cannot be adequately measured, EC50 values for immobilization or loss of equilibrium or both and TLm values must be based on 96-hr experiments. Animals excluded from this category include air-breathing molluscs, unionid clams, operculate snails, and bivalve molluscs that can "close up" and thus prevent exposure to the toxicant.

For each species for which satisfactory studies with a particular toxicant have been performed the acute concentration (either TLm or EC50) is calculated as the geometric mean of the values obtained in all flow-through tests in which the toxicant concentrations were measured. If no such tests are available, the acute concentration is calculated from the geometric mean of all available acute values (*i.e.*, values based on flow-through tests in which the toxicant concentration was not measured and static and renewal tests in which the initial total toxicant concentration is specified). For example, Table 2.9 lists 96-hr TLm values for fathead minnows (Pimephales promelas) exposed to dieldrin.

Once mean acute concentrations have been determined for the appropriate number and variety of species, the values are ordered from lowest to highest and the common logarithm (to base 10) of each value is calculated. Beginning with the lowest logarithm, the logarithms are then assigned to intervals or cells, each of which has a width of 0.11 log units. The factor of 0.11 is an estimate of the average precision of the mean acute values and is equivalent to a factor of $10^{0.11}=1.29$ in the acute concentration. For example, if the lowest mean acute value were A. then all mean acute values between A and 1.29 A would be assigned to the first cell, all mean values between 1.29A and $(1.29)^2$ A would be assigned to the second cell, and so forth.

TABLE 2.9

Dieldrin 96-Hr TLm Values for Fathead Minnous (Pimephales Promelas)*

Chemical Source	*TLm (ppb)*
90% Dieldrin	18
90% Dieldrin	18
85% Dieldrin	36
85% Dieldrin	24
85% Dieldrin	16
85% Dieldrin	25
85% Dieldrin	23

*All experiments were conducted in a static mode. The dieldrin concentration is based on the amount of dieldrin added to the water; actual concentrations were not measured.

The calculation is illustrated for the dieldrin saltwater TLm data in Table 2.10. The table includes results from a total of 21 species, and satisfies the EPA conditions for both diversity and number of organisms. It is noteworthy that the TLm values vary by more than a factor of 70. The lowest TLm is 0.7 ppb for the pink shrimp Penaeus duorarum and the highest TLm is 50.0 ppb for the grass shrimp Palaemonetes vulgaris. A cumulative proportion distribution for the TLm values is

TABLE 2.10

Dieldrin Species Mean Acute Values in Saltwater*

Species	*Mean acute value (ppb)*
Grass shrimp, Palaemonetes vulgaris	50.0
Northern puffer, Sphaeroides maculatus	34.0
Eastern oyster, Crassostrea virginica	31.2
Striped mullet, Mugil cephalus	23.0
Striped bass, Morone saxatilis	19.7
Hermit crab, Pagurus longicarpus	18.0
Threespine stickleback, Gasterosteous aculatus	14.2
Korean shrimp, Palaemon macrodactylus	10.8
Sheepshead minnow, Cyprinodon variegatus	10.0
Mummichog, Fundulus heteroclitus	8.9
Grass shrimp, Palaemonetes pugio	8.6
Sand shrimp, Crangon septemspinosa	7.0
Bluehead, Thalassoma bifasciatum	6.0
Striped Killifish, Fundulus majalis	5.0
Atlantic silverside, Menidia menidia	5.0
Mysid shrimp, Mysidopsis bahia	4.5
Dwarf perch, Micrometrus minimus	3.5
Shiner perch, Cyamatogaster aggregata	2.3
Chinook salmon, Oncorhynchus tshawytscha	1.5
American eel, Anguilla rostrata	0.9
Pink shrimp, Penaeus duorarum	0.7

*From EPA (1980b).[3]

shown in Fig. 2.6. The TLm values have been assigned to cells according to the EPA procedures. The first cell extends from 0.7 ppb to (0.7) (1.29)=0.90 ppb, while the last (17th) cell extends from (0.7) $(1.29)^{16}$=40.2 ppb to (0.7) $(1.29)^{17}$=51.9 ppb. In fact, however, only the first two nonempty intervals are used in calculating the final acute value. The procedure is as follows. First, the geometric mean of the TLm values in the first two nonempty cells is calculated, In this case both the first and second cell contain only one value, and hence there is no need to calculate geometric means. The cumulative proportions for the first and second cells are 1/21 =0.048 and 2/21 =0.95. The procedure for calculating the final acute value is now to interpolate (in this case) or extrapolate the TLm values to a cumulative frequency of 0.05. The calculation is based on a linear fit to the log transformed TLm data. For the first cell the log of the mean TLm is log (0.7)= —0.155, and for the second cell log (0.9)= —0.046. The corresponding proportions are 0.048 and 0.095, respectively. A straight line through these two points follows the equation log (TLm)= —0.26+2.3

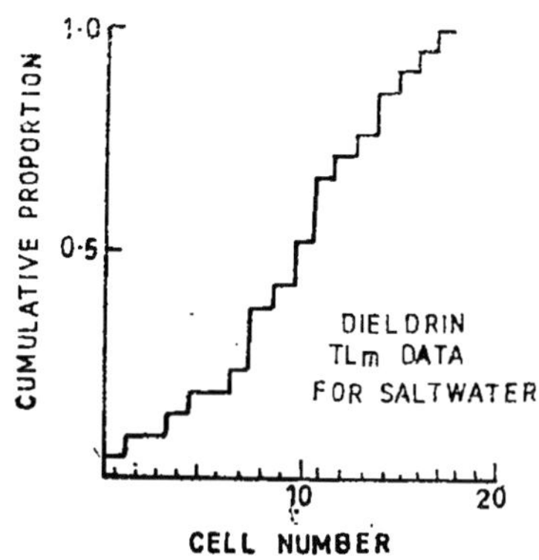

Fig. 2.6: Cumulative frequency distribution of saltwater TLm data for dieldrin. The abscissa is effectively a logarithmic scale, with the first cell containing values between 0.7 and 0.9 ppb and the last (17th) cell containing values between 40.2 and 51.9 ppb. [Data from EPA (1980).[2]]

(P), where P is the cumulative proportion associated with the TLm. If P=0.05, then log (TLm)=−0.15 and TLm=0.71 ppb. Based on the data available for dieldrin, it is expected that if a large random sample of saltwater species were taken, only 5 per cent would have TLm values less than 0.71 ppb. The final saltwater acute value for dieldrin established by the EPA in 1980[2] is therfore 0.71 ppb.

IV. CHRONIC TOXICITY DETERMINATION

There are basically two approaches to determining chronic values of a toxicant. The first approach is similar to the method used for determining acute values and requires a similar number and variety of test organisms. The second approach is to divide the final acute value by an average lethal/sublethal or acute/chronic ratio. We consider the former approach first.

A. Duration of Experiments

Although acute toxicity studies can be completed in 48 to 96 hr, chronic toxicity studies require much longer periods of time, and the lengths of exposure must in general be appropriate to the species. According to 1980 EPA guidelines,[1] the experiments must be performed in flow-through systems (renewal systems are acceptable for daphnids) in which the toxicant concentrations are measured in the test solution. The 1980 EPA guidelines also impose the following constraints on chronic toxicity studies.[1]

1. Life cycle chronic toxicity tests should consist "of exposures of each of several groups of individuals of a species to a different concentration of the toxicant throughout a life cycle." The tests must begin with embryos or newly hatched young less than 48 hr old (less than 24 hr old for daphnids) and should continue through maturation and reproduction. For fish, the exposure should end not less than 24 days (90 days for salmonids) after hatching of the next generation. For daphnids the study should include an analysis of

survival and growth of both young and adults, maturation of males and females, embryot spawned per female, embryo viability (salmonids only), and hatchability.

2. Partial life cycle tests are acceptable with fish species that require more than a year to reach sexual maturity. It should be possible to expose all major life stages to the toxicant and still complete the study within 15 months. As in the case of the complete life cycle studies, each of several groups of individuals of the species should be exposed to a different concentration of the toxicant. Exposure should begin with immature juveniles at least 2 months prior to active gonad development and should end not less than 24 days (90 days for salmonids) after the hatching of the next generation. The data analysis should be similar to that used in the complete life-cycle studies.

3. Early life stage toxicity tests may be used if complete or partial life cycle tests with the same species are unavailable. The exposures should begin shortly after fertilization and should last 28 to 32 days (60 days posthatch for salmonids) through early juvenile development. Survival and growth data should be analyzed.

B. Calculation of the Chronic Value

As an example of the calculation of a chronic value for a toxicant, let us consider a hypothetical experiment on species X in which the effect of copper on the number of embryos spawned per female is being tested. The control group of organisms is placed in water containing no added copper, while groups of the remaining organisms are placed in water containing various concentrations of copper. The number of embryos spawned by individuals in each group is recorded, and the results for each group compared to the results of the control group. The effect of a given copper concentration is considered to be statistically significant if the number of embryos spawned per female in the control group and the given test group differ at $p=0.05$. The results of the hypothetical experiment are shown in Table 2.11. The chronic value is calculated as "the geometric mean of the lower and upper chronic limits from a chronic test."[1] The lower chronic limit is the highest tested concentration that produced an effect not significantly different from that of the control group and below which no tested concentration produced a statistically significant effect. In this case the lower chronic limit would be 0.05 ppm. The upper chronic limit is the lowest tested concentration that produced an effect significantly different from that of the control group and above which all tested concentrations produced statistically significant effects. In this case the upper chronic limit would be 2.00 ppm. The chronic value with respect to the effect of copper on embryos produced per female by species X would therefore be $[(0.05)(2.00)]^{1/2}=0.32$ ppm.

TABLE 2.11

Results of Hypothetical Experiment to Determine Effects of Copper on Embryo Production by Females of Species X[1]

Copper concentration (ppm)	*Embryo production per female different from control at p=0.45*
0.005	No
0.01	No
0.05	No
0.10	Yes
0.20	No
0.50	Yes
1.00	No
2.00	Yes
5.00	Yes
10.00	Yes

[1]Chronic values $=[(0.05)(2.00)]^{1/2}=0.32$ ppm.

The mean chronic value for a species is obtained by simply calculating the geometric mean of all the chronic values for the species. Sample calculations based on data for the Cladoceran Daphnia magna are shown in Table 2.12. If these data were the only information available on the chronic toxicity of these compounds to D, magna, the mean chronic values would be as indicated in the last column of Table 2.12. Since only two effects were studied, the mean value is simply the square root of the product of the chronic values in the previous two columns. Note that since species will invariably be more sensitive with respect to some kinds of chronic effects than to others, the types of chronic effects actually studied can influence to a certain extent the mean chronic value. It is in part for this reason that EPA guideline request that data be collected and analyzed on a variety of chronic effects. Once mean chronic values have been obtained for a suitable number and variety of species, the final chronic value for the given toxicant is calculated in the same manner as the procedure for calculating the final acute value (*e.g.*, Fig. 2.6).

TABLE 2.12

Chronic Toxicity of Four Compounds to Daphnia magna[1]

Compound	*Chronic values (ppb)*		
	Survival of adults	*Reproduction*	*Mean chronic value*
1, 1, 2-Trichloroethane	32,000	18,000	24,000
Dieldrin	100	32	57
Pentachlorophenol	180	320	240
3, 4-Dichloroaniline	56	5.6	18

[1]Data from Adema, D.M.M. (1978). Daphnia magna as a test animal in acute and chronic toxicity tests. Hydrobiol, 59, 125-134.

C. Acute/Chronic Ratios

The second approach for calculating final chronic values is to make use of acute/chronic ratios. This procedure may be used if chronic toxicity information is not available for the number and variety of organisms required to calculate a final chronic value directly, but if both acute and chronic toxicity values are available for at least three species subject to the following constraints:

1. at least one is a fish;
2. at least one is an invertebrate; and
3. at least one is a freshwater species (if the ratio is being applied to freshwater data) or at least one is a saltwater species (if the ratio is being applied to saltwater data).

An acute/chronic ratio is calculated from a particular study in the following manner. The chronic value for the particular effect being studied is calculated as the geometric mean of the upper and lower chronic limits as previously described (Table 2.11). The acute value used in calculating the ratio may be obtained in several ways. Preferably the acute value is taken to be the "arithmetic average of the results of all standard flow-through acute tests in which the concentrations were measured and which are from the same study as the chronic test."[1] If such acute tests were not performed in conjunction with the chronic studies, one may take the acute, value to be the result obtained in a standard acute test performed at the same laboratory and with the same species, toxicant, and dilution water. If neither of these two types of data is available, the acute value may simply be taken to be the species mean acute value. The species mean acute/chronic ratio is then calculated as the geometric mean of all acute/chronic ratios available for that species.

The final acute/chronic ratio for the given toxicant may be calculated in one of two ways. If there is no major trend apparent among the acute/chronic ratios for the different species and the species mean acute/chronic ratios lie within a factor of ten for a number of species, the final acute/chronic ratio is calculated as the geometric mean of all the species mean acute chronic ratios for both saltwater and freshwater species. However, if the species mean acute/chronic ratios seem to be positively correlated with the species mean acute values, the final acute/chronic ratio should be taken to be the acute/chronic ratio of species whose acute values lie close to the final acute value as previously calculated (Fig. 2.6). The final chronic value for the toxicant is then calculated by dividing the final acute value by the final acute/chronic ratio.

An example of the use of acute/chronic ratios is shown in Table 2.13 for the pesticide dieldrin. Chronic values for this pesticide were available for only four species in 1980 when the guidelines for dieldrin were established, and therefore it was necessary to use acute/chronic ratios to establish the final chronic value. Acute toxicity values were available in only three of the four cases where chronic effects were studied, but fortunately the three species studies satisfied the conditions for calculating acute/chronic ratios in both freshwater and saltwater. Since the calculated acute/chronic ratios differ by less than a factor of two, it is appropriate to calculate the final acute/chronic ratio for dieldrin by simply taking the geometric mean of the three separate ratios, which is $[(11)\,(9.1)\,(6,2)]^{1/3}=8.5$. Final acute values for dieldrin in freshwater and saltwater were

TABLE 2.13

Acute/Chronic Ratios for Dieldrin Toxicity to Three Species of Aquatic Animals

Species	*Acute value (ppb)*	*Chronic value (ppb)*	*(Ratio)*
Rainbow trout, Salmo qairdneri	2.5	0.22	11
Guppy, Poecilia reticulata	4.1	0.45	9.1
Mysid shrimp, Mysidopsis bahia	4.5	0.73	6.2

2.5 and 0.71 ppb, respectively.[2] Hence, the final chronic values for freshwater and saltwater are 2.5/8.5=0.29 ppb and 0.71/8.5=0.084 ppb, respectively.

V. TOXICITY TO PLANTS

The guidelines for calculating acute and chronic concentrations of a toxicant clearly pertain to aquatic animals. Nevertheless aquatic plants may be adversely affected by the presence of toxicants in the water, and sometimes at exceedingly low concentrations (Fig. 2.1). According so EPA guidelines, a substance is toxic to a plant at a given concentration if the growth of the plant is decreased in a 96-hr or longer experiment with an alga or in a chronic test with a vascular plant. Growth may be measured in a variety of ways, including photosynthetic rate, change in dry weight, or change in chlorophyll concentration. Table 2.14 provides a summary of data on the toxicity of dieldrin to aquatic plants. In this particular case, the concentrations that reduced plant growth were substantially higher than both the final chronic and final acute values for animals.

TABLE 2.14

Concentrations of Dieldrin that Reduce Growth by Aquatic Plants

Species	*Effects*	*Concentration (ppb)*
Scenedesmus quadricaudata	22 per cent reduction in biomass in 10 days	100
Navicula seminulum	50 per cent reduction in growth in 5 days	12,800
Wolffia papulifera	Reduced population growth in 12 days	10,000
Agmenellum quadruplication	Reduced growth fate	950

and would therefore not influence the final water quality guidelines for dieldrin. However, because plants are the chief producers of organic matter in most aquatic food clains, it is critical that the effects of toxicants on aquatic plants be assessed. Reductions in photosynthetic rates and/or changes in the composition of the plant community may cause repercussions throughout the entire food web. The so-called final plant value is the lowest concentration of the toxicant that reduces plant growth in an appropriate experiment. In the case of dieldrin, the final plant value would be 100 ppb in freshwater and 950 ppb in saltwater.

VI. RESIDUE VALUES

Up to this point we have considered the direct adverse effects on plants and animals caused by toxicants present in the water. However, once these toxicants are incorporated into the organisms themselves, there is obviously a potential for other kinds of adverse effects caused by the transfer of the toxicants through the food web. One obvious concern is that the toxicants may be transfered to humans as the result of consumption of contaminated fish or selfish. The rationale in 1980 EPA guidelines behind establishing a final residue value is actually twofold; first, to "prevent commercially or recreationally important aquatic organisms from exceeding relevant FDA [Food and Drug Administration] action levels," and second, to "protect wildlife, including fishes, and birds, that eat aquatic organisms from demonstrated adverse effets."[1] The residue value, which is the maximum permissable concentration in the water based on the foregoing two considerations, is calculated by dividing the maximum permissable tissue concentration by an appropriate bioconcentration factor (BCF), where the BCF "is the quotient of the concentration of a substance in all or part of an aquatic organism divided by the concentration in water to which the organisms has been exposed."[1] The maximum permissable tissue concentration is the lower of (1) a FDA action level for fish oil or the edible portion of fish or selfish and (2) the tissue concentration that results in "a maximum acceptable dietary intake based on observations on survival, growth or reproduction in a chronic wildlife feeding study."[1] BCF values based on experimental laboratory tests are acceptable only if the concentration of the toxicant in the water was actually measured and if the concentration in the tissues of the test organism reached a steady-state value. A steady-state value in this context is defined as a concentration that did not change significantly over a period of time, "such as two days or 16 per cent of the length of the exposure, whatever is longer." If steady state was not reached after a period of 28 days, it is permissable to calculate the BCF based on the longest exposure over 28 days.

In certain respects BCF values based on field data would be preferable to BCF values based on laboratory experiments because some pathways of exposure may exist in the field but not in the laboratory. For example, a fish placed in an aquarium tank may take up a toxicant directly from the water. However, a fish in the natural environment may also assimilate the toxicant in the food it eats. Because of such considerations, 1980 EPA guidelines indicate that BCF values based on field studies should be used in preference to BCF values based on laboratory studies for a given toxicant. However, data from field exposures are acceptable only if "it is known that the concentration of the substance was reasonably constant for a long enough period of time over the range of territory inhabited by the organisms."[2]

Special considerations apply to lipid soluble toxicants. First, the percentage of lipids should be measured in the tissue for which the BCF is calculated. A normalized BCF is then calculated by dividing the unnormalized BCF by the percentage of lipids. In effect this normalized BCF is the BCF that one would expect to measure if the tissue contained 1 per cent lipids. The geometric mean of all the normalized BCFs is then calculated using data from both freshwater and saltwater organisms. A residue value is then calculated by dividing the maximum permissable tissue concentration by the product of the mean normalized BCF and the percentage of lipids appropriate to the particular kind of tissue. The appropriate percentage of lipids is 15 per cent, 16 per cent, and 100 per cent for the edible portions of freshwater fish, the edible portions of saltwater fish, and fish oil, respectively. In the case of chronic feeding studies with wildlife, the appropriate percentage of lipids according to the EPA is "the per cent lipids of an aquatic species or group of aquatic species which constitute a major portion of the diet of the wildlife species."[1]

The following example illustrates how the final residue value was determined for dieldrin, a lipid soluble substance. Table 2.15 lists the BCF values reported for dieldrin in various freshwater and saltwater species. Only two studies actually reported the percentage of lipids. The geometric mean of the normalized BCF values for these two species is $[(1160/1)(2300/1.1)]^{1/2}=1557$. Since the FDA action level for dieldrin in fish, shellfish, and fish oil is 0.3 ppm on a wet tissue basis, residue values would be calculated as follows:

freshwater fish:

$$(0.3 \text{ ppm}/[(1557)(15)]=0.013 \text{ ppb}$$

saltwater fish:

$$(0.3 \text{ ppm})/(1557)(16)]=0.012 \text{ ppb}$$

fish oil:

$$(0.3 \text{ ppm})/(1557)(100)]=0.0019 \text{ ppb}$$

Two additional residue values may be calculated based on the BCFs for dieldrin in the edible portions of fish and selfish. The highest BCF for the edible portion of a freshwater fish in 2993 for the channel catfish (2.15). The corresponding residue value would therefore be (0.3 ppm/2993)=0.10 ppb. The highest BCF value for the edible portion of a saltwater species is 8000 for the eastern oyster (Table 2.15), and the corresponding residue value would be (0.3 ppm/8000)=0.038 ppb.

Only two aquatic wildlife feeding studies have been performed with dieldrin, and both involved rainbow trout (Salmo gairdneri). Studies show that a dieldrin concentration of 0.36 ppm in the diet of rainbow trout alters the fish's ability to detoxify ammonia, and a diet containing 0.36 ppm dieldrin

TABLE 2.15

Bioconcentration Values in Freshwater Species of Deildrin on a Wet Weight Basis

Species	*Tissue*	*% Lipid*	*BCF*
Alga, Scenedesmus obliquus	—	—	128
Community dominated by the alga Tribonema minus	—	—	5,558
Community of algae including Stigeoclonium subsecundum Synedria ulna, Epithemia sorex, Cocconeis placentula var. euglypta, and Nitzschia sp.	—	—	3,188
Cladoceran, Daphnia magna	—	—	1,395
Freshwater mussel, Lampsilis siliquoidea	Whole body	1	1,160
Steelhead trout (newly hatched alevin), Salmo gairdneri	Whole animal	—	3,225
Lake trout (yearling), Salvelinus namaycush	Whole body	—	68,286
Channel catfish, Ictalurus punctatus	Dorsal muscle	—	2,385
Channel catfish, Ictalurus punctatus	Dorsal muscle	—	2,993
Guppy, Poecilia reticulata	Whole animal	—	12,708
Guppy, Poecilia reticulata	Whole animal	—	28,408
Eastern oyster, Crassostrea virginica	Edible tissue	—	8,000
Crab, Leptodinus floridanus	Whole body	—	400
Sailfin molly, Poecilia latipinna	Edible tissue	—	4,867
Spot, Leiostomus xanthurus	Whole body	1.1	2,300

alters the fish's metabolism of phenylalanine. Although the lipid content of the diet was not reported, it is obvious that the lipid content could not have exceeded 100 per cent, and therefore a minimum residue value based on these two feeding studies would be (0.36 ppm)/[(1557) (100)]=0.0023 ppb. This figure is still larger than the residue value of 0.0019 ppb based on the fish oil action level. In fact, the smallest of any of the residue values is 0.0019 ppb, and that figure therefore becomes the final residue value for both freshwater and saltwater. It is noteworthy, however, that this final residue value may be too high because when a residue value is calculated from an FDA action level, "On the average half of the individuals of the species would have concentrations above the FDA action level." Furthermore, "the results of many chronic feeding studies are concentrations that cause adverse effects."[1]

VII. THE TWO-NUMBER CRITERION

Present EPA water quality guidelines are based on a two-number criterion. The first number is the concentration of the toxicant that should not be exceeded at any time or place. The second number is the level that the average concentration of the toxicant should not exceed over any 24-hr period. The rationale for using a two-number criterion is that organisms can tolerate brief exposures to toxicant concentrations substantially higher than can be tolerated over a longer period. This fact is evident in Fig 2.4. There is invariably a negative correlation between the concentration of a toxicant that causes a particular adverse effect and the duration of exposure. Under the present system four toxicant concentrations are utilized in establishing the two-number criterion. These four concentrations are the final acute value, the final chronic value, the final plant value, and the final residue value. In the foregoing discussion we have seen in some detail how these values are determined from toxicological studies. The four values for dieldrin in freshwater and in saltwater are shown in Table 2.16. The concentration that should not be exceeded at any time is simply equated to the final acute value, which is 2.5 and 0.71 ppb in freshwater and saltwater, respectively. The level that the 24-hr average concentration should not exceed is the smallest of the final chronic value, the final plant value, and the final residue value. In both freshwater and saltwater, the final residue value of 0.0019 ppb is the smallest of these three numbers and therefore becomes the second number in the two-number criterion.

TABLE 2.16

The Four Toxicant Concentrations used to Establish Water Quality Guidelines for Dieldrin in Freshwater and Saltwater[1]

	Value in freshwater (ppb)	*Value in saltwater (ppb)*
Final acute value	2.5	0.71
Final chronic value	0.29	0.084
Final plant value	100	950
Final residue value	0.0019	0.0019
Concentration to be exceeded at no time	2.5	0.71
Value the 24-hr average concentration should not exceed	0.0019	0.0019

[1] From EPA (1980b).[3]

VIII. COMPLICATING FACTORS

A. Interactions with Harmless Substances or Conditions

The toxicity of a particular substance to aquatic organisms may sometimes depend strongly on the concentrations of other substances or on environmental conditions that by themselves are harmless. For example, in aquatic systems a continuous interconversion takes place between ammonium ions (NH_4^+) and un-ionized ammonia (NH_3). The equation describing this interchange can be written $NH_3+H^+ \rightleftharpoons NH_4^+$.

Although NH_4^+ is nontoxic, NH_3 is highly toxic. The values for NH_3 96-hr TLm for a variety of freshwater fish have been found by the EPA to be in the range 0.2=0.9 ppm. The equilibrium distribution of NH_3 and NH_4^+ is obviously a function of pH, and as the pH is raised (H^+ activity reduced), the equilibrium shifts toward NH_3. At steady state the expression $[NH_3]\,[H^+]/[NH_4^+]$ equals $10^{9.3}$, and hence at a pH of 9.3, the concentrations of NH_3 and NH_4^+ are equal. Obviously whenever the pH is increased by one unit, the ratio $[NH_3]/[NH_4^+]$ increases by a factor of 10. Thus, water that was virtually harmless at a pH of 7 might become extremely toxic if the pH were raised to 9 (as a result, for example, of the removal of CO_2 due to photosynthetic uptake) as a result of the conversion of NH_4^+ to NH_3. Because of this consideration, it is futile in terms of water quality to assay merely for NH_3+ NH_4^+ (by standard procedures) without simultaneously measuring the pH to determine the equilibrium concentration of NH_3. Similarly, 1972 FPA water quality guidelines[3] for ammonia are written in terms of un-ionized ammonia (NH_3) rather than NH_3=NH_4^-.

Temperature is another factor that can greatly influence the toxicity of a substance without itself causing any significant stress. In general terms the performance of an organism is positively correlated with temperature below the optimum temperature for the organism, and then rapidly drops as the temperature increases further (Fig. 2.7). The rate of increase of the respiration rate or metabolic rate of an organism is usually expressed in terms of the organism's Q_{10}, which is the change in respiration rate or metabolic rate per 10°C increase in temperature. For temperature conformers (*i.e.*, cold-blooded organisms) metabolic rates typically increase by a factor of 2 to 3 for every 10°C rise in temperature as long as the temperature is not too close to the lethal temperature for the species. In other words, typical Q_{10} values for cold blooded organisms are 2 to 3. Obviously any increase in respiration rate implies an increased demand for oxygen, and as a result, the TLm for oxygen can be expected to be positively correlated with temperature. This fact in part underlies the tendency of fish kills caused by inadequate oxygen concentrations to occur during the summer when water temperatures are highest. An increase in the respiration rate of a fish obviously implies an increase in the rate at which water is pumped over the gills. As a result, the intensity of exposure of the fish to any toxic substance in the water is increased. The effects of a given toxicant concentration on the organism may therefore become apparent more rapidly in warm water than in cold water.

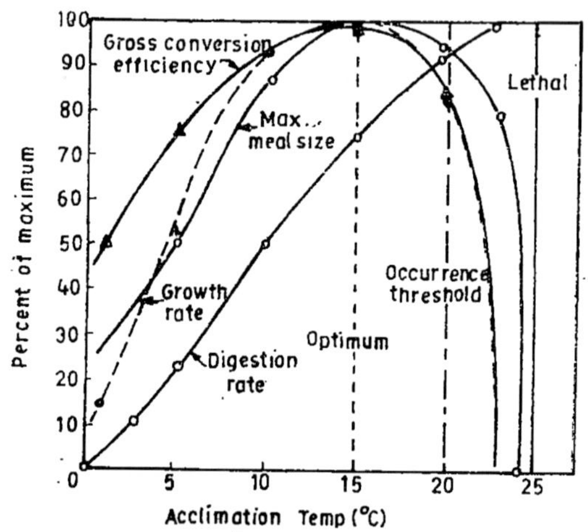

Fig. 2.7. Performance of sockeye salmon (Oncorhynchus nerka) as a function of acclimation temperature. [Reprinted from Brett, J. R. (1971). Energetic responses of salmon to temperature. A study of some thermal relations in the physiology and fresh water ecology of sockeye salmon (Oncorhynchus nerka). Am. Zool. 11, 99—113. ©American Society of Zoologists.]

Figure 2.4 provides another clear example of the interaction between otherwise harmless water quality characteristics and toxicant effects. The toxicity of zinc to rainbow trout is clearly correlated in a negative way with water hardness. Similar correlations between metal toxicity and water hardness have been noted for a variety of toxic metals. The mechanism responsible for the correlation is not well understood, and considering the wide range of metals for which the effect has been observed, it is unlikely that exactly the same mechanism is operative in all cases. Water hardness, which is strictly speaking a measure of the concentration of calcium and magnesium ions in the water, tends to be correlated with a number of other water quality characteristics such as pH and alkalinity. It is therefore possible that at least in some cases the observed negative correlation between metal toxicity and water hardness actually reflects a cause-and-effect relationship with one of these other covariates.

1. Incorporation into Water Quality Guidelines

How can one account for these interactions between toxicant effects and water quality characteristics in establishing water quality criteria? EPA guidlines do make allowance for such interactions, and at least in the case of some heavy metals, these interactions have explicitly been taken into acount in formulating the guidelines. Allowance for the interactions may be made if enough data are available to show that the toxicity of a substance to two or more species is similarly affected by the given water quality characteristic. For establishing the final acute equation the procedure is as follows.

First, for each species for which comparable acute toxicity data are available at two or more values of the water quality characteristic, a least squares regression is performed on the logtransformed data. Transformations other than the logarthmic transformation may be used if a better fit to the data is obtained. Meaningful regression slopes must be obtained from the data for a least two species of justify further use of the equations.

If at least two such meaningful slopes are obtained, the arithmetic average of all the meaningful slopes is calculated. Then for each species, an intercept A is calculated from the equation

$$Y = A + bX \qquad \ldots(1)$$

where b is the previously calculated arithmetic average slope and X and Y are the means of the transformed values of the water quality characteristics and acute toxicity values for the given species, respectively. For each species the mean acute intercept is then calculated by back transforming A. For example, if the original data were in fact transformed using common logarithms, then the mean acute intercept for each species would be 10^A. The final acute intercept is then calculated using the same procedures as used to calculate the final acute value (*e.g.*, Fig. 2.6). If the original data were log transformed, the final acute equation has the form

$$\text{acute value} = a(\text{water quality characteristic})^b$$

where a is the final acute intercept.

Application to cadmium

As an example of the application of this procedure, we consider the freshwater water quality guidelines for cadmium. As is the case for many other heavy metals, the toxicity of cadmium is negatively correlated with water hardness. This fact is apparent from an examination of Fig. 2.8, which is a log-log plot of TLm values for seven freshwater species versus water hardness. Although only two or three data points were available in the case of several species, the range of water hardness values is sufficiently great in all cases to allow a "meaningful" slope to be calculated. The arithmetic mean of the seven slopes is 1.05 ± 0.39. Therefore b is set equal to 10.5. At the time EPA water quality guidelines for cadmium were established, data on water hardness and the acute toxicity of cadmium were available for a total of 29 freshwater species. For each of these 29 species, the mean of the log-transformed values of water hardness (X) and TLm values (Y) were used in Eq. (1) to calculate values of A. The back-transformed values of $A(=10^4)$ are listed along with the 29 species in Table 2.17, and a cumulative frequency distribution of these mean acute intercepts analogous to Fig. 2.6 is shown in Fig 2.9. The data in Table 2.17 and Fig. 2.9 were then analyzed by a method identical to that employed with the dieldrin TLm data in Table 2.10 and Fig. 2.6 to yield a final acute intercept of 0.024 ppb. Thus, for fresh water organisms the first criterion in the EPA guidleines is that the cadmium concentration (in ppb) not exceed 0.024 (water hardness)$^{1.05}$.

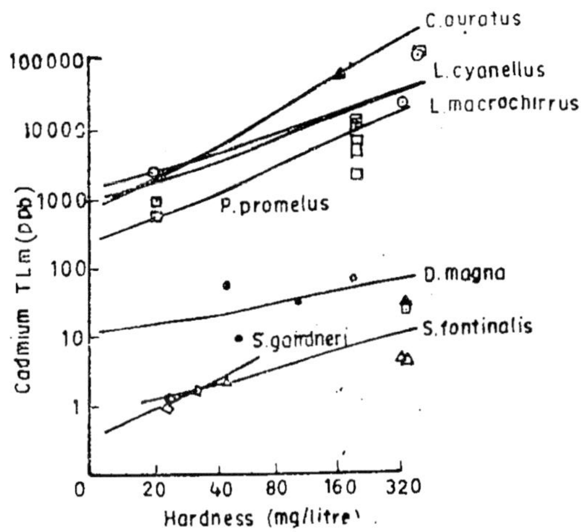

Fig. 2.8. Cadmium freshwater TLm values as a function of water hardness for seven species. Both scales are logarithmic. [Data from EPA (1980c).[4]]

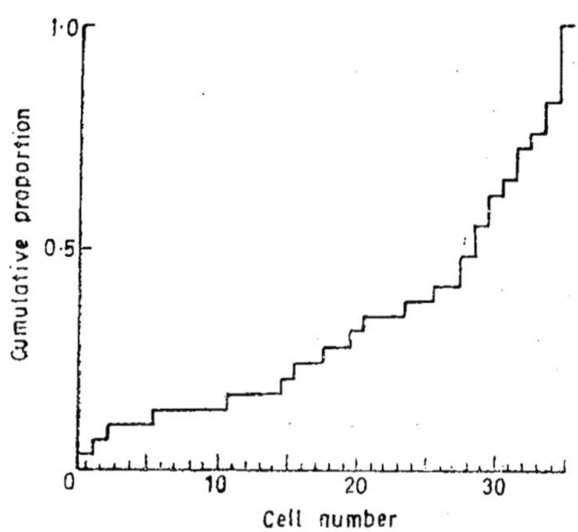

Fig. 2.9. Cumulative frequency distribution of freshwater acute toxicity intercepts for cadmium. The abscissa is effectively a logarithmic scale, with the first cell containing values between 0.02 and 0.026 ppb, and the last (35th) cell containing values between 110 and 142 ppb. Data from Table 2.17.

TABLE 2.17

Cadmium Mean Acute Intercepts for 29 Freshwater Species[1]

Species	*Mean acute intercept (ppb)*
Snail, Amnicola sp.	138
Mosquitofish, Gambusia affinis	135
Goldfish, Carassius auratus	134
Damselfly, unidentified	133
White perch, Morone americanus	125
Green sunfish, Lepomis cyanellus	91.4
Three spine stickleback, Gasterosteus aculeatus	86.7
Bluegill, Lepomis macrochirus	80.7
Caddisfly, unidentified	55.9
Guppy, Lebistes returulatus	54.7
Flagfish, Jordanella floridae	47.0
Fathead minnow, Pimephales promelas	38.2
Northern squawfish, Ptychochellus oregonensis	35.9
Mayfly, Ephemerella grandis grandis	30.3
Bristle worm, Nais sp.	28.0
Pumpkinseed, Lepomis gibbosus	22.3
Midge, Chironomus	19.7
American eel, Anquilla rostrata	12.2
Rotifer, Philodina acuticornis	7.01
Carp, Cyprinus carpio	3.57
Snail, Physa gyrina	2.87
Banded killifish, Fundulus diaphanus	1.67
Scud, Gammarus sp.	1.15
Cladoceran, Simocephalus serrulatus	0.87
Caldoceran, Daphnia magna	0.29
Chinook salmon, Oncorhynchus tshawytscha	0.09
Rainbow trout, Salmo qairdneri	0.04
Brook trout, Salvelinus fontinalis	0.03
Striped bass, Morone saxitalis	0.02

[1]From EPA (1980c).[4]

There are two possible techniques for determining the final chronic equation. The first procedure is directly analogous to the method used to determine the final acute equation, except that a meaningful chronic slope need be obtained for only one species. The second approach it to obtain an acute/chronic ratio by the methods previously described, and to simply divide the final acute intercept by this ratio to obtain a final chronic equation. We again use cadmium as an example.

At the time the EPA water quality guidelines were written, the data needed to establish the relationship between water hardness and the chronic toxicity of cadmium were available for only three species, the cladoceran Daphnia magna, brook trout (Salvelinus fontinalis), and channel catfish (Ictalurus punctatus). The data are shown graphically in Fig. 2.10. The chronic toxicity of cadmium to channel catfish was almost completely independent of water hardness, but for brook trout the slope of the log-log plot was 1.01, almost identical to the mean acute slope from the acute toxicity results. The slope

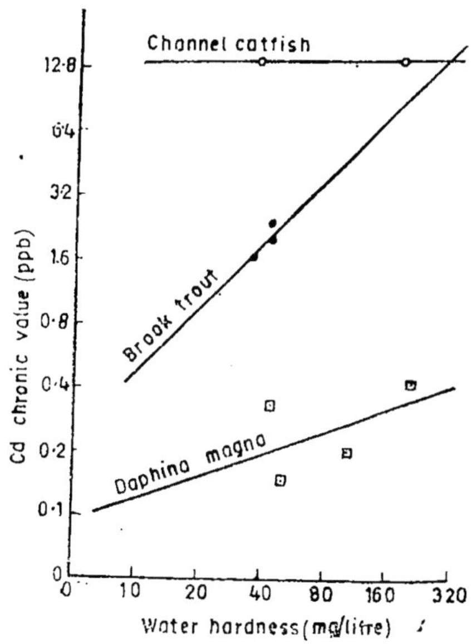

Fig. 2.10: Chronic values of cadmium for three fresh-water species as a function of water hardness. [Data from EPA (1980c).[4]]

of a log-log fit to the four D. magna data points was 0.35. However, three of the D. magna data points are from a single study, while the fourth is from a second study. If only the three data points from the single study are included, the slope of the log-log fit is 0.79. Based on these results, the EPA concluded, "Hardness affects the acute and chronic toxicity of cadmium similarly."[4] This statement seems to be an oversimplication considering the fact that data were available for only three species and that the calculated log-log slopes ranged from 0.03 to 1.01. Parhaps a more correct statement would have been that there were too few data and too much scatter in the calculated slopes to allow a meaningful comparision, but at least the null hypothesis that the acute and chronic slopes are the same cannot be rejected based on the available information.

Acute/chronic ratios for cadmium were available for a total of four freshwater species and one saltwater species. The data are shown in Table 2.18. Only the data for D. magne permit an examination of whether the acute/chronic ratio is correlated with water hardness. For the four D. magna data points, the correlation coefficient between the acute/chronic ratio and water hardness is small (—0.19) and statistically insignificant, a result consistent with the assumption that acute and chronic toxicity vary in a similar way with water hardness. Although the acute/chronic ratios do not vary widely for cadmium, there is a positive correlation (R=0.79) between the mean acute intercepts and the mean acute/chronic ratios for the four species in Table 2.18. It therefore does not seem appropriate to use a simple averaging technique to determine the final acute/chronic ratio. Since D. magna is clearly the most sensitive of the species tested, the results for this species were used to determine the final acute/chronic ratio. The geometric mean of the four D. magna acute/chronic ratios is 122. The final chronic intercept for freshwater therefore becomes 0.024 ppb, and the final acute intercept, divided by 122, is 0.000197 ppb. The final acute chronic equation for Cd therefore becomes (in ppb) 0.000197 (water hardness)$^{1.05}$. Ultimately this equation became the second criterion in the two-number freshwater water quality guidelines for cadmium because neither effects on aquatic plants nor biological concentration implied a more strict criterion.

TABLE 2.18

Cadmium Mean Acute Intercepts and Acute/Chronic Ratios for Four Freshwater Species[1]

Species	*Water hardness (mg/litre)*	*Mean acute intercept (ppb)*	*Acute/chronic ratio*
Bluegill, Lepomis macrochirus	207	80.7	422
Flag fish, Jordanella floridae	44	47.0	431
Fathead minnow, Pimephales promelas	201	38.2	130
Cladoceran, Daphnia magna	45	0.29	191
	53	—	66
	103	—	157
	209	—	111

[1]EPA (1980c).[4]

B. Conditioning and Acclimation

The toxicity of a particular stress to an aquatic organism can depend very much on the previous life history of the organism. This reflects the ability of many, organisms to acclimate or adapt to changes in the environment and in particular to the presence of toxic substanees in the water. Figure 2.11 provides a good example of the effects of adaptation on the toxic effects of temperature stress. The data are taken them a study of both lethal and sublethal temperature stresses on young sockeye salmon. Two features of the graph are of particular interest. First, sublethal effects on spawning and growth become apparent outside a much narrower temperature range than is the case for lethal effects. For example, the fish do not begin to die outright until the temperature exceeds 21-25°C or falls below 5-7°C. These observations simply illustrate and re-emphasize the importance of distinguishing chronic effects from lethal or acute effects. Second, the range of temperatures within which the fish can effectively carry out certain activities is shifted to higher temperatures as the acclimation temperature is increased. For example, a young sockeye salmon acclimated to a temperature of 20°C would evidently die if placed in water at 4°C, but a similar fish acclimated at 10°C would not die if placed in the same water. There is, however, a limit to the adaptive capabilities of aquatic organisms. For example, it is apparent from Fig. 2.11 that young sockeye salmon cannot spawn outside the approximate temperature range 5-14°C, regardless of the acclimation temperature.

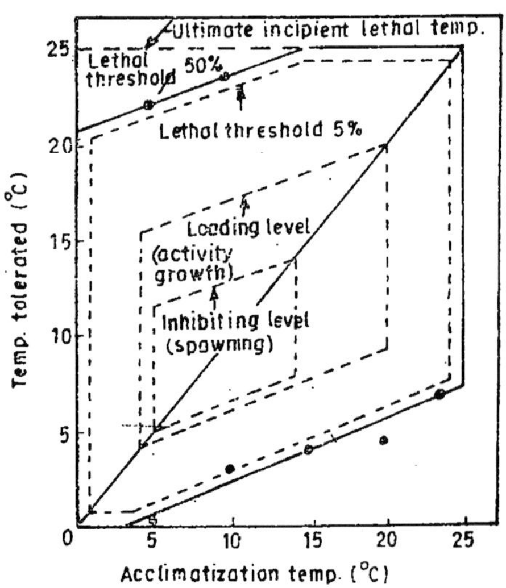

Fig. 2.11: Temperature tolerance zones for young sockeye salmon (Oncorhynchus nerka) as a function of acclimation temperature. [Reprinted with permission from Brett, J. R. (1960). Thermal requirements of fish—three decades of study. In "Biological Problems of Water Pollution," (C. M. Tarzwell, ed.), pp. 110-117. U. S. Department of Health, Education, and Welfare.]

In some cases exposure of an organism to a stress may actually increase the sensitivity of the organism to that stress, but in many cases prolonged exposure to a sublethal level of stress may actually desensitize or acclimate the organism to the stress. For example, many fish are capable of acclimation to low oxygen levels as well as to high and low temperatures. Some fish have been shown to acclimate to ammonia, cyanide, pH, phenol, synthetic detergents, and zinc. In organisms such as bacteria that can multiply rapidly, adaptation may actually involve genetic changes. Mutant strains may by chance be more resistant to a particular stress than the other members of the population and over time will gradually dominate the population. Genetic selection of this sort is believed to have been responsible for the emergence of pesticide-resistant strains of about 200 different insect pests. Among organisms that reproduce more slowly genetic selection would take much longer, and acclimation in the short term must therefore involve other mechanisms such as biochemical, physiological, or behavioural adjustments. The processes involved in these adjustments are in general not well understood, but probably involve changes in the functioning of hormones and/or enzymes and the response characteristics of the organism's nervous system.

In a general sense organisms that adapt to a stress can be classified as being either "conformers" or "regulators." For example, cold-blooded animals are conformers with respect to temperature because their internal body temperature is strongly correlated with the temperature of the environment, Warm-blooded animals are regulators with respect to temperature because their internal temperature is relatively independent of the temperature of the environment. Of course, for both conformers and regulators there is an optimum internal state at which the organism functions most effectively. Acclimation among regulators usually produces a change in the range of the external variable within which the organism can effectively control its internal state, but does not change the optimum state of the organism. Acclimation among conformers may involve a change in the organisms's optimal internal state and usually involves a change in the range of internal variability within which the organism can function efficiently.

Should the possible effects of acclimation be considered when setting water quality guidelines? The attitude of the EPA is that acclimation effects should in general not be considered and that toxicity data obtained with acclimated organisms should not be used in deriving water quality guidelines. The 1980 EPA guidelines rationale is that acclimation is the exception rather than the rule, and that, "Rarely, if ever, can acclimation be depended on to protect organisms in a field situation because concentrations often fluctuate and motile organisms do not stay in one location very long." However, "Data obtained with acclimated organisms may be acceptable for use in deriving some site-specific criteria."[1]

C. Interactions Between Toxic Substances

The discussion to this point has considered the effect of a toxicant as if it were the only stress-producing substance. However, what effects can be anticipated if two or more toxic substances are present in the water at the same time? For example, according to the present EPA guidelines concentrations of lead and copper should not exceed 3.8 and 22 ppb, respectively.[1] Now suppose that a particular body of water is found to contain 2.5 ppb lead and 15 ppb copper. Would this water be likely to exert a chronic stress that would in some sense violate the intentions of the EPA guidelines?

The answer to this question is that the effect of two or more toxic substances on the organisms in the water will depend on the manner in which the toxicants interact. The following example illustrates the various ways in which two toxicants might interact. Suppose that the 96-hr TLm for a particular organism is 1.0 and 10 ppb for toxicants A and B, respectively. Now suppose that toxicants A and B are added to the water in a flow-through acute toxicity experiment so that the concentrations of A and B are in fact 1.0 and 10 ppb, respectively. The test organisms are placed in the water and their survival monitored for 96 hr. Any of the following outcomes are possible.

1. Exactly half the organisms are dead after 96 hr. This result is exactly what one would have expected if one had added 1.0 ppb A but no B to the water, or 10 ppb B but no A. In this case one says that there has been no interaction between A and B.

2. Fewer than half the organisms are dead after 96 hr. In this case the toxicity of the mixture is apparently less than the toxicity of a solution containing only 1.0 ppb A or only 10 ppb B, and the interaction between A and B is said to be antagonistic.

3. If more than half the organisms are dead after 96 hr, there are several possible interpretations.

   (a) If the percentage of dead organisms is exactly what would have been expected if the water contained only 2.0 ppb A or only 20 ppb B, then the interaction of A and B is said to be strickly additive. In other words, 1 ppb A produces the same toxicity as 10 ppb B, and when the two substances are present together the toxicity of the mixture can be calculated by assuming that every 10 ppb of B are equivalent to 1.0 ppb of A.

   (b) If the fraction of organisms that are dead after 96 hr is greater than in case a, then the interaction is said to be supraadditive.

   (c) If the percentage of organisms dead after 96 hr is greater than 50 per cent but less than in case a, then the interaction is said to be infraadditive.

It is useful in discussing these interactions to speak in terms of toxic units rather than actual concentrations. In this hypothetical example, one acute toxic unit of A would be 1.0 ppb A, and one acute toxic unit of B would be 10 ppb B. The possible interactions between A and B may then be indicated graphically, as in Fig. 2.12.

How should one take account of these interactions in setting water quality guidelines? One obvious practical problem is that there is not a great deal known about the interactions of toxic substances. In fact, a review of current EPA water quality guidelines[1] shows that in many cases insufficient information is available to establish numerical guidelines for toxicants cansidered separately. In fact, the present guidelines simply make no allowance for interactions between toxicants. In effect, this policy amounts to assuming that in all cases there is no interaction between toxicants (figure 2.12). In response to criticisms concerning this policy, the EPA has stated. "Synergism and antagonism are possible between numerous combination [sic] of two or more pollutants, and some data indicate that such interactions are not only species specific, but also vary with the ratios and absolute concentrations of the pollutants and the life stage of the species."

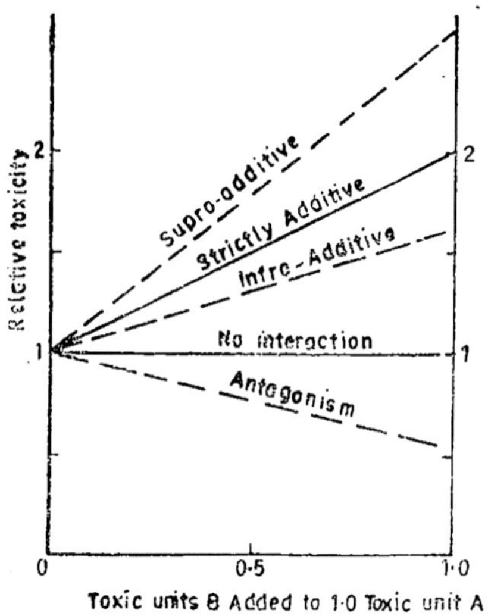

Fig. 2.12: Relative toxicity of a mixture of toxicants A and B under various assumptions concerning the interaction of the two toxicants.

In other words, not much is known, the interactions may be very complicated, and given these facts it is very hard to estimate the toxicity of a solution containing more than one toxic substance. However, reference to Fig. 2.12 shows that in two cases, calculation of the toxicity of a mixture of toxicants is straightforward. First, if there is no interaction between the toxicants, then the toxicity of the mixture would be determined by the toxicant present in the greatest number of toxicity units. For example, if a solution contained 0.75 toxic units of A and 0.5 toxic units of B, then if A and B do not interact, the toxicity of the solution would be equal to that of a solution containing only 0.75 toxic units of A. Similarly, if the interaction were strictly additive, the toxicity of the mixture would be equal to that of a solution containing only 1.25 toxic units of A or only 1.25 toxic units of B. In other words, if one assumes that there is no interaction or that the interactions are strictly additive, then the toxicity of a mixture of toxicants can easily be calculated. However, if the interactions are supra-additive, infraadditive, or antagonistic, then one must have detailed information about the nature of the interaction to calculate the toxicity of the mixture.

It is apparent from Fig. 2.12 that the assumption of no interaction is conservative only if the true interactions are antogonistic. The assumption of no interaction would underestimate the true interactions if the true interactions were in any sense additive. However, the assumption of strictly additive interactions would be conservative in all cases except the case in which the true interactions were supraadditive. Thus, if the goal of setting water quality standards were to err on the side of safety, it would make more sense to assume that toxicants interact in a strictly additive manner than to assume that they do not interact at all.

There is no doubt that some toxicants do not interact in a strictly additive manner. For example, the toxicity of cadmium is decreased by other metal ions, particularly low concentrations of copper and zinc, and it is known that the cholinesterase-inhibiting pesticides EPN (O-ethyl-O-paranitrophenyl phenylphosphonothioate) and Malathion interact in a supra-additive manner. The TLm of a mixture containing equal numbers of toxic units of EPN and Malathion is only one twelfth the TLm of the individual compounds. The effect is due to the ability of EPN to inhibit the enzyme system that is responsible for detoxifying Malathion after it has been converted to Malaoxon. In contrast, chlorinated hydrocarbons such as DDT that induce some of the liver microsomial enzyme systems may interact antagonistically with chemicals that are detoxified by the same enzyme systems (*e.g.*, hexabarbital). Although it is admittedly illogical to expect the toxicity of poisons with different toxicological properties and different concentration—response curves to be strictly additive, "Nevertheless, the method has been found to work empirically...If, for the present, the poisons can be regarded as agents producing stress...each of which produces a degree of shock with resulting non-specific effects, it might be considered reasonable that summation of the overall stress is possible." [From Brown, V. M. (1980). The calculation of the acute toxicity of mixtures of poisons to rainbow trout. Water Res. 2, 723-733.]

Application of the strictly additive interaction assumption to water quality guidelines was made in the 1972 EPA water quality criteria, where it was noted.

> The system of additing different toxicants in this way is based on the premise that their lethal actions are additive. Unlikely as it seems, this simple rule has been found to govern the combined lethal action of many pairs and mixtures of quite dissimilar toxicants, such as copper and ammonia, and zinc and phenol in the laboratory...The rule holds true in field studies...The method of addition is useful and reasonably accurate for predicting thresholds of lethal effects in mixtures. [From EPA (1972),[3] p. 122.]

Application of the addition principle is illustrated by the data in Table 2.19 for a hypothetical body of water. Acute toxicity units are calculated by dividing the concentration of each toxicant by its corresponding final acute value. The acute toxicity units are then simply added (strictly additive assumption) to determine the toxicity of the mixture. If the sum of the acute toxicity units is 1.0 or greater, the mixture would be considered to exert an acutely toxic stress, a condition that would be unacceptable at any time. However, according to the EPA (1972),[3]

> There is also evidence of a lower limit for additive lethal effects. For ammonia and certain other pollutants, levels below 0.1 of the lethal concentration do not seem to contribute to the lethal action of a mixture...This lower cutoff point of 0.1 of the LC50 should be used when it is necessary to assess the lethal effects of a mixture of toxicants.

Applying this reasoning to the data in Table 2.19, one would ignore the acute toxicity units for cadmium, chromium copper, silver, and zinc because they are all less than 0.1. One would add the toxicity units contributed by lead and selenium to obtain a sum of 0.75 acute toxicity units. Since this figure is less than 0.1, the water would presumably not exert an acutely toxic stress.

TABLE 2.19

Example of Calculation of Acute Toxicity Units in Water Containing Various Heavy Metals and With a Hardness of 100 mg/litre

Metal	*Final acute value (ppb)*	*Concentration (ppb)*	*Acute toxicity units*
Cd	3.0	0.12	0.04
Cr	4700	188	0.04
Cu	22	1.76	0.08
Pb	170	85	0.50
Se	260	65	0.25
Ag	4.1	0.205	0.05
Zn	320	16	0.05

A similar procedure would be used to calculate chronic toxicity units. The number of chronic toxicity units contributed by a particular substance would equal the concentration of the substance divided by the EPA final chronic value. The chronic toxicity units for each toxicant would then be added, and if the 24-hr average of the sum of the chronic toxicity units exceeded 1.0, the water would be considered to exert a significant chronic stress. Once again, however, there is reason to believe that chronic toxicity units smaller than a certain value should be ignored when summing the contributions from each toxicant. In 1972 the EPA recommended that chronic toxicity units less than 0.2 be ignored, but the recommendation was admittedly based on relatively little information. Since the concept of additive interactions has not been incorporated into the present EPA guidelines, the rationale for a threshold level of acute or chronic toxicity units has not been reinvestigated. Ultimately it would seem that some allowance must be made for toxicant interactions. If a general guideline in establishing water quality criteria is to err on the side of safety and if detailed information on the interactions between combinations of toxicants is not available, then the assumption of strictly additive interactions would seem preferable to the assumption of no interactions at all.

IX. PUBLIC HEALTH

In the previous discussion we have considered the use of toxicological studies for purposes of setting water quality guidelines primarily for the protection of marine and freshwater aquatic organisms. An equally if not more important use of toxicology studies is to establish water quality guidelines for the protection of the general public. Two general categories of effects are considered under the present EPA guidelines. These categories are carcinogenic effects and toxic effects other than carcinogenic effects. We consider the use of toxicology data to establish water quality guidelines in these two categories in the following discussion.

A. Carcinogenic Effects

A crucial assumption in the water quality guidelines for carcinogenic substances is that there is no threshold level of exposure below which there is no adverse effect. The water quality guidelines are then written simply in terms of the additional lifetime risk (probability) of contracting cancer as a result of consumption of water containing a certain concentration of the toxicant, and from consumption of a specified amount of fish and shellfish taken from the same water. A substance is assumed to be carcinogenic to humans if it has been shown to induce a satistically significant carcinogenic response in animals in experimental studies or in humans through accidental exposure. Once it has been determined that a given substances is carcinogenic to humans, an effort is made to calculate the water concentration that is estimated to cause a lifetime carcinogenic risk of 10^{-5}. This risk is the additional probability over background that a person would contract cancer sometime in his or her lifetime as a result of continuous exposure to the compound. The point of view is taken that the carcinogenic process involves irreversible damage to DNA, a position that is supported by the fact that a large proportion of agents that cause cancer are also mutagenic.

1. The Dose—Response Model

There is reason to expect that the biological response characteristic of mutagenesis is associated with a linear nonthreshold dose-response relationship at low dose rates. The actual model used by the EPA has the form

$$A(d) = 1 - \exp(-q_1 d - q_2 d^2 - \ldots - q_k d^k) \qquad \ldots(2)$$

where A(d) is the extra risk over background associated with a dose d, and the q_k are constants to be determined by a least squares curvefitting procedure. At low values of d, Eq. (2) reduces to the form

$$A(d) = q_1 d \qquad \ldots(3)$$

and therefore satisfies the condition of being a linear nonthreshold model at low doses. The procedure is to fit Eq. (2) to experimental data relating dose d to the incidence of cancer in experimental animals and to then calculate an upper 95 per cent confidence limit on the value of q_1. This upper 95 per cent confidence limit on q_1 is referred to as q_1^*. A lower 95 per cent confidence limit on the dose associated with an extra lifetime cancer risk of 10^{-5} would therefore be $10^{-5}/q_1^*$.

There are two important considerations concerning the calculation of d and q_1. First, it is assumed that milligrams per surface area per day is an equivalent measure of dose between species. Since surface area is approximately proportional to body weight raised to the two-third power, it is customary and acceptable to calculate dose as $mg/W^{2/3}/day$, where W is the average weight of the animals. Thus, in a given experiment dose would be calculated from the following equation:

$$d = l_e m / L_e W^{2/3} \qquad \ldots(4)$$

where L_e is the duration of the experiment, l_e the duration of exposure, and m the average daily dose (in mg) during the administration of the agent.

Second, if the duration of the experiment was less than the natural lifespan of the test animal, then q_1^* (or more generally the polynomial $q_i d^i$) is increased by multiplying by a factor $(L/L_e)^3$, where L is the natural life span of the test animal. The rationale for this procedure (from the 1980 EPA guidelines) is that, "The age specific rates for humans increase at least by the 2nd power of the age and often by a considerably higher power...Thus, we would expect the cumulative tumor rate to increase by at least the 3rd power of age."[1]

The calculation of the concentration in the water corresponding to a given level of risk is then calculated on the assumption that a person consumes 2 litres of water and 6.5 g of fish and shellfish per day on average. The amount of the toxicant consumed each day is therefore C(2+0.0065R), where C is the concentration of the toxicant (in mg/litre) in the water and R the bioconcentration factor between fish or shellfish and water. On the assumption that the average person weighs about 70 kg (154 lb), the daily dose associated with the lifetime cancer risk of 10^{-5} would be $70^{2/3}$ and $10^{-5}/q_1^*$, where q_1^* has units of $kg^{2/3}/d/mg/$ Equating this expression to C(2+0.0065R) and solving for C gives

$$C = \frac{70^{2/3} \times 10^{-5}}{q_1^*(2+0.0065R)} \qquad \ldots(5)$$

The bioconcentration factor R is calculated in a manner similar to that used previously in the calculation of final residue values for the protection of aquatic species. However, different assumptions are made regarding toxicants soluble in lipids. In the calculation of final residuce values, a figure of 15 per cent lipids was used for freshwater fish and 16 per cent for salt-water fish. These figures are in fact representative of freshwater lake trout and saltwater Atlantic herring, both of which are fatty species. The rationale for using fatty species as a standard in determining final residue values is that FDA action levels are applied on a species-by-species basis and therefore the final residue value must be chosen low enough to protect even the fatty species. However, a calculation indicates that the lipid content of fish and shellfish consumed by the average person is only about 3 per cent, and it is this latter figure that is used in establishing the bioconcentration factor R.

The following example illustrates how the value of C was determined in the case of dieldrin. Figure 2.13 shows results of a study of the hepatocellular carcinoma response of female CF-1 mice given various concentrations of dieldrin continuously in their diet. When the data were analyzed using the procedures previously described, the value of q^*_1 was found to be 7.37 $kg^{2/3}/d/mg$. The bioconcentration factor previously calculated for dieldrin was 1557 normalized to 1 per cent lipid. Since the fish and shellfish consumed by the average person are assumed to contain 3 per cent lipid, the value of R is 3(1557)=4671. The value of C calculated from Eq. (5) is therefore 0.071×10^{-5} mg/litre=0.71 ng/litre. If a similar calculation is made for the consumption of aquatic organisms only, excluding the consumption of water, then the factor of 2 in the denominator of Eq. (5) is ignored and the calculated value of C is 0.76 ng/litre.

B. Noncarcinogenic Effects

In the case of noncarcinogenic substances, the water quality

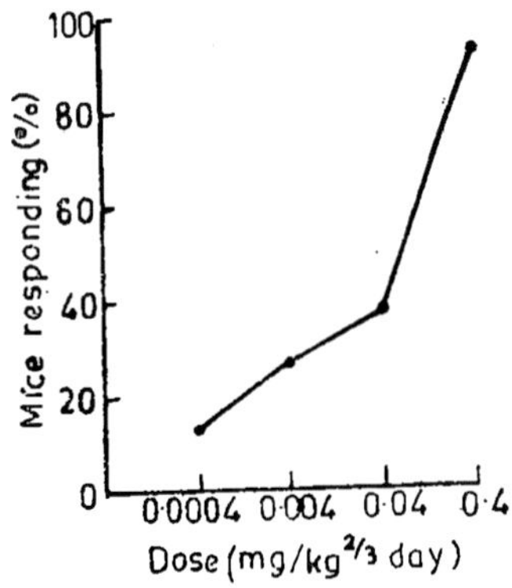

Fig. 2.13: Incidence of hepatocellular carcinoma in female CF-1 mice as a function of dieldrin dose. [Data from Walker, A.I.T., Thorpe, E., and Stevenson, D.E. (1972). The toxicology of dieldrin (HEOD). Long-term oral toxicity studies in mice. Food Cosmet. Toxicol. 11, 415-432.]

guidelines are based on the assumption that there is in fact a threshold concentration below which there are no adverse effects. That such an attitude must be taken in the case of at least some substances is obvious because, according to the 1980 EPA guidelines, "Some chemicals, particularly certain metals, are essential to biological organisms at low levels but may be toxic and/or carcinogenic at high levels."[1] The determination of threshold levels may be based on noncarcinogenic response levels of the following types:

NOEL: no observed effect level

NOAEL: no observed adverse effect level

LOEL: lowest observed effect level

LOAEL: lowest observed adverse effect level

These response levels are listed in order of degree of conservatism with respect to establishing the threshold level, with the NOEL leading to the most conservative (most protective) guideline and the LOAEL to the least conservative guidelines.

According to EPA precedures,[1] a single NOEL is not acceptable for establishing the threshold level, and if multiple NOELs are available without additional data on NOAELs, LOELs, or LOAELs, then the highest NOEL should be used in deriving the water quality guideline. The NOALs, LOELs, and LOAELs may be used singly or in combination to establish the threshold level. Uncertainty factors or safety factors are applied to the response levels according to the criteria listed in Table 2.20. An additional judgemental uncertainly factor ranging from 1 to 10 must be applied to LOAEL data. Once an acceptable daily intake has been established by application of the uncertainty factor,

TABLE 2.20

Uncertainty factors for deriving criteria for threshold effect toxicants from NOEL, NOAEL, LOEL, and/or LOAEL data[1]

Nature of experimental results	*Uncertainty factor*
Valid experimental results from studies on prolonged ingestion by humans with no indication of carcinogenicity.	10
Experimental results of studies of human ingestion not available or scanty (*e.g.*, acute exposure only) with valid results of long-term feeding studies on experimental animals, or in the absence of human studies, valid animal studies on one or more species; no indication of carcinogenicity.	100
No long-term or acute human data; scanty results on experimental animals with no indication of carcinogenicity.	1000

[1]From National Academy of Sciences (1977). "Drinking Water and Health." National Academy of Sciences, Washington, D.C.

the criterion (concentration in water) for the toxicant is calculated from the equation

$$C=(ADI-DT-IN)/(2+0.0065R) \qquad \ldots(6)$$

where C is the criterion, ADI the acceptable daily intake, DT is the estimated nonfish dietary intake, IN the estimated daily intake due to inhalation, and R the bioconcentration factor (in litres/kg). Equation (6) assumes that a person consumes an average of 2 litres of water and 6.5 g of fish and shellfish per day. It is recognized that adequate information on DT and IN may not always be available, in which case an assumption must be made concerning total exposure. Lack of information about or variability in DT and IN obviously can create a degree of uncertainty in the derived water quality guidelines.

Application of the foregoing methodology is illustrated in the following example of cadmium. In this case there are a number of examples of human beings exposed to levels which clearly caused adverse effects. In areas of Japan where itai-itai disease is prevalent, it is known that about 85 per cent of the cadmium intake is due to the consumption of cadmium-contaminated rice. The disease is associated with tubular proteinuria, the incidence of which rises above that observed in control populations when the cadmium concentration in the rice exceeds about 0.45 ppm. Japanese in the area are known to consume about 430 g/day of rice. Therefore, the LOEL for Japanese is (430)(0.45)/0.85=228 μg/day. For Americans, and western Europeans, who are somewhat larger than Japanese, the corresponding LOEL would be expected to be about 301 μg/day. Since only about 5 per cent of ingested cadmium is actually absorbed by the body, the figure of 301 μg/day ingested corresponds to only 15 μg/day absorbed.

Similar conclusions have been reached from other studies. For example, a study of workers who inhaled cadmium dust concluded that effects began to appear when the concentration of cadmium in the air equaled 21 μg/m^3. If one assumes that the average person inhales about 10 m^3 of air per day, that lung retention of cadmium is about 25 per cent efficient, and that the workers were exposed to the contaminated air 5 days per week, the calculated level of absorption corresponding to the appearance of effects was (21)(10)(0.25)(5/7) = 38 μg/day. Similarly, the Working Groups of Experts for the Commission of European Communities has estimated that the threshold level of cadmium exposure by ingestion is about 200 μg/day, which would correspond to an absorption of 10 μg/day.

Applying the uncertainty factor of 10 in Table 2.20 appropriate for studies on prolonged ingestion by humans, one would conclude that an ADI would be about 1.0-3.8 μg/day for cadmium. It is apparent from Table 2.21 that even for a nonsmoker, the average rate of cadmium absorption exceeds exceeds the lower bound of the ADI range. However, it is also apparent that drinking water contributes a very minor fraction of the total cadmium absorbed. This fact arises because the average drinking water in the United States contains only 1.3 ppb cadmium. In fact, only 3 of a total of 969 U.S. community water supplies studied by the EPA contained cadmium in excess of the existing drinking water standard of 10 ppb.

TABLE 2.21

Average Daily Absorption of Cadmium by Americans[1]

Exposure source	*Exposure*	*Cd Intake* (μg)	*% Absorbed*	*Absorption* (μg)
Air, ambient	0.03 $\mu g/m_3$	0.6	25	0.15
Air, smoking (one pack)	3.0 μg/pack	3.0	25	0.75
Food	—	30.0	5	1.50
Drinking water	1.3 μg/litre	2.6	5	0.13

[1]From EPA (1980c).[4]

How much cadmium would an average person absorb from drinking water and the consumption of fish and shellfish if all water contained 10 ppb cadmium. Bioconcentration factors

(BCFs) for cadmium in some commercially important fish and shellfish are shown in Table 2.22. The BCFs are consistently higher for the five bivalve molluscs than for the other two species. The geometric mean BCFs for the bivalve molluscs and

TABLE 2.22

Cadmium Geometric Mean Species BCFs for Consumed Species[1]

Species	*Mean BCF*
Brook trout, Salvelinus fontinalis	16.5
American oyster, Crassostrea virginica	898
Soft shell clam, Mya arenaria	160
Quahog, Mercenaria mercenaria	83
Bay scallop, Aquipecter irradians	585
Common mussel, Mytilus edults	113
Crab, Carcinus maenas	5.9

[1]From EPA (1980c).[4]

the two other species are 240 and 10, respectively. Of the 6.5 g/day of fish and shellfish consumed by the average American, only 0.8 g is accounted for by bivalve molluscs. Freshwater and estuarine fish and shellfish account for the other 5.7 g/day. It therefore seems appropriate to weight the effective BCF according to the relative amounts of the different kinds of fish and shellfish, and the weighted BCF is therefore [(0.8)(240)+(5.7(10)]/6.5=38. The daily intake of cadmium from drinking water and the consumption of fish and shellfish if all water contained 10 ppb cadmium would therefore be [2+0.0065 (38)](10)=22.5 μg/day. If absorption of this consumed cadmium is 5 per cent efficient (Table 2.22), the daily absorption would be 1.1 μg/day. Of this figure, only about 0.1 μg/day comes from the consumption of fish and shellfish. Regardless of the actual concentration of cadmium in the water, consumption of fish and shellfish will clearly be of minor importance compared to drinking water as a source of cadmium absorption to the average American.

Based on similar calculations, the EPA in the 1980 guidelines concluded, "It appears that a water criterion needs to be no more stringent than the existing primary drinking water standard (10 μg/litre) to provide ample protection of human health." This conclusion seems debatable. If a drinking water supply actually contained 10 μg/litre of cadmium, absorption from drinking water alone for the typical person would average 1.0 μg/day. If the typical absorption of 1.5 μg/day from food and 0.15 μg/day from air are added to this figure (Table 2.21), the total absorption is 2.65 μg/day, which is over twice the lower bound of the ADIs for cadmium. The EPA's conclusion seems to have been based largely on the fact that, "Water constitutes only a relatively minor portion of man's daily cadmium intake."[4] However, this fact arises not because average water contains 10 ppb cadmium but because average water contains almost eight times less cadmium. A downward revision in the water quality guidelines for cadmium therefore seems warranted.

X. COMMENTARY

At the present time there are approximately 30,000-50,000 chemicals in use or being distributed through the environment, and an additional 700-1000 are added each year. Although not all these chemical are toxic, merely keeping up-to-date information on the possible toxicity of so many compounds is a prodigious task. The foregoing discussion has indicated how involved the process of setting water quality guidelines is for even a single chemical. Large amounts of the right kind of information are needed. When the EPA finally published water quality guidelines in 1980 for 64 of the 65 pollutants for which Congress had demanded guidelines in 1977, the lack of pertinent information was immediately apparent. Of the 65 chemicals, a complete set of numerical criteria based on the EPA's guidelines were available for only 8 of them: aldrin/dieldrin, cadmium, chlordane, endosulfan, endrin, heptachlor, nickel, and selenium. This fact dramatically illustrates the nature of the problem that confronts governmental agencies charged with the responsibility of controlling water pollution. On the one hand, there is an increasingly urgent demand to establish water quality guidelines for the large number of toxic chemicals that pollute both our groundwater and surface water. On the other hand, there is a need to develop intelligent and legally defensible techniques for establishing those guidelines. Ultimately the government agencies must decide in a somewhat arbitrary manner how much of what kind of toxicologic information is necessary and sufficient to establish water quality criteria. Establishing the rules of the game is obviously a critical first step, and now that the first step has been taken, information gaps will hopefully begin to be filled at a rapid rate. Invariably there will remain disagreement over whether the methodology repuired to establish the guidelines is too simple or too cumbersome, too strict or too lax, and too detailed or too ambiguous. There will remain differences of opinion as to what types of toxicological data are useful and acceptable and how best to utilize those data in developing guidelines. The wisdom and collaboration of many conscientious scientists and administrators will be needed if the legal efforts to control the quality of our water is to keep pace with the growing threat of pollution.

BIBLIOGRAPHY

Epstein, S.S., Brown, L.O., and Pope. C. (1982). "Hazardous Waste in America." Sierra Club, San Francisco.

Laws, E.A. (1981). "Aquatic Pollution." Wiley (Interscience), New York.

Loomis, T.A. (1974). "Essentials of Toxicology," 2nd ed. Lea and Febriger, Philadelphia.

Pye, V.I., Patrick, R., and Quarles, J. "Groundwater Contamination in the United States." University of Pennsylvania Press, Philadelphia.

Warren, C.E. (1971). "Biology and Water Pollution Control." Saunders, Philadelphia.

Zwick, D., and Benstock, M. (1971). "Water Wasteland." Grossman, New York.

NOTES

1. Environmental Protection Agency (1980). "Water Quality Criteria Document; Availability." Fed. Reg. 45 (231), 79318-79379. Referred to in text as EPA (1980a).
2. Environmental Protection Agency (1980), "Ambient Water Quality Criteria for Aldrin/Dieldrin, EPA 440/5-80-019. Referred to in text as EPA (1980b).
3. Environmental Protection Agency (1972). "Water Quality Criteria." EPA/R3/73/033. Referred to in text as EPA (1972).
4. Environmental Protection Agency (1980). "Ambient Water Criteria for Cadmium." EPA 440/5-80-025. Referred to in text as EPA (1980c).

3

WATER RESOURCE SYSTEMS

There are three categories of systems of concern to water resources management. First is the natural water system, which is controlled by the physical laws of nature. Second, there are the institutional systems that regulate the use and control of water according to laws and regulations established by society. Third, there are the user systems constructed for the purpose of utilizing the waters of the natural system in accordance with the constraints imposed by the institutional systems.

A. Basic Elements of the System

Any attempt to characterize nature's water system must start with the hydrologic cycle. The latter represents the continuous movement of water above, on, and below the earth's surface. There is no specific starting point for this cycle, but it is convenient to begin with the water in the atmosphere. This vapour, when condensed, falls as precipitation on land and ocean surfaces. Eventually the water returns to the atmosphere via the combined efforts of evaporation and transpiration. This elemental circuit, however, is a gross simplification of the complex routes followed by different water droplets.

Precipitation that falls on the land can be subdivided into three major categories: water that moves across the surface and enters streams, water that is held in surface storage of one form or another, and water that penetrates or infiltrates the earth's surface. [See *Soil Physics*.]

The water that when striking the earth's surface begins to flow across the surface is initially termed overland flow. This plane surface flow eventually infiltrates the surface or reaches a channeled section. Once into the channel system the ultimate terminus is the ocean. An exception to this direct route can occur when the channelized flow depth exceeds adjacent groundwater levels. In this case some of the flow will move into the saturated groundwater or into temporary bank storage.

Water temporarily stored on the surface of vegetation and/or in small, relatively impermeable depressions is said to be intercepted storage. The bulk of this storage is subsequently evaporated. More significant portions of surface storage occur in large natural depressions, lakes, marshes, snowfields, and stream channels, and in manmade impoundments, This storage is eventually released via evaporation and subsurface seepage or by both natural and regulated discharge to the channel system.

Water that penetrates the earth's surface is said to have infiltrated. Some of this infiltrated water is retained near the surface as soil moisture and is subsequently returned to the atmosphere via the process of evapotranspiration. If the soil moisture storage increases to field capacity, that is, the maximum amount of moisture the soil can hold by capillarity, gravitational flow begins. This downward movement of water is termed percolation. If this flow strikes an impervious layer it will move laterally and possibly exit the subsurface by seepage into a ditch or ravine and become part of the surface flow. This portion of the surface discharge is termed interflow. In the absence of the aforementioned impervious diversion, the percolation will continue downward until it reaches the gravity groundwater table. Groundwater flow then proceeds down gradient within the saturated aquifer. In many cases this groundwater flow will eventually discharge into a stream. This part of streamflow is often referred to as the baseflow. Alternatively, some groundwater by virtue of its location discharges directly to the ocean.

B. Gross Allocations Within the System

In a global sense the total amount of freshwater is essentially constant. Table 3.1 provides a typical example of the estimated global distribution of freshwater. The volume of saltwater in the oceans is 15 times greater than the total supply of freshwater.

TABLE 3.1

Approximate Distribution of Freshwater[1]

Classification	*Volume[2] (cu miles)*	*Per cent of total*
Ice (glaciers)	5,800,000	84.945
Groundwater	960,000	14.158
Lakes and reservoirs	37,000	0.549
Soil moisture	20,000	0.294
Atmospheric vapour	3,400	0.049
Rivers and streams	300	0.004
	6,820,700	100.000

[1]Heath, R. C. (1982). Basic Ground-water Hydrology, U.S. Geological Survey Water Supply Paper 2220 p. 1.
[2]1 cu mile equals 4.16 km^3.

Published estimates suggest the average daily precipitation in the conterminous United States approximates 4200 billion gal and represents about 10 per cent of the atmospheric moisture that crosses this area. On average, an amount exceeding two-thirds of the annually precipitated water returns to the atmosphere via natural processes of evaporation and transpiration. Streamflow discharge to the oceans approximates 28 per cent of the precipitation and an additional 3 per cent reaches these same water bodies as subsurface flow. These average values can be misleading. Difficulty arises due to the lack of uniform areal coverage and to the absence of steady-state conditions. For example, in parts of the High Plains as much as 98 per cent of the incoming precipitation will eventually exit the region by the evapotranspiration process. Yet, in portions of the Gulf Coast no more than 50 per cent of the precipitation is devoted to fulfilling evapotranspiration requirements.

Temporal variations in the relative allocations are sizable for any catchment. In the long-term the percentage of rainfall that becomes surplus to the evapotranspiration process is largely a function of climate. In the short-term the local geology, human culture, and individual storm characteristics all serve to modify the day-to day split between surface flow, subsurface flow, and evapotranspiration.

To illustrate, the basic hydrologic mass balance equation for any natural catchment, or segment thereof, and for some specified duration of time can be expressed as follows:

$$P=R+E+\Delta S \quad \ldots(1)$$

where P is the precipitation, R the runoff, both surface and subsurface, E the evapotranspiration, and ΔS the change in storage. All terms are expressed in equivalent basinwide units of depth. In the long-term, and presuming the absence of subsurface mining, the ΔS term should approach zero. In such a case the mass balance reduces to

$$P=R+E \quad \ldots(2)$$

Dividing each term by P results in the following expression for the ratio of runoff to precipitation.

$$R/P=1-E/P \quad \ldots(3)$$

Equation (3) supports the prior comment that in the long-term this ratio is primarily a function of climatic characteristics.

In the short-term the ΔS term in Eq. (1) becomes most important. In this case the ratio of runoff to precipitation is materially impacted by geological characteristics, man induced cultural characteristics, antecedent moisture conditions and individual storm characteristics. Geological and cultural characteristics can vary significantly within a given catchment and among catchments. Storm characteristics and antecedent moisture conditions may vary widely from storm to storm. Accordingly, one is presented with an infinite array of potential boundary conditions, and the analysis of the physical response in nature's water system becomes most challenging.

C. Quality Considerations

Water quality is often more important than water quantity. Contrary to views shared by some segments of the lay public, the quality of nature's water system is highly variable. Explanation for much of this natural variability can be found in well-known principles of the chemistry of aqueous solutions. However, water quality is a complex issue and continues to become ever more so due to the actions of mankind.

Water has long been recognized as an outstanding solvent. This is nowhere better demonstrated than in the natural water system. Many water quality concerns stem from the reactions that occur at the interface between water and rock. Rapid advances in the science of geochemistry during the past several decades have served to explain many of the early mysteries associated with the natural variance in water quality. In a less scientific sense most people can accept the concept that mineralization of the flow will increase with the length and duration of subsurface percolation. The rationale that oceans and land-locked lakes that receive mineralized discharges will become salty is also easily accepted. However, the processes by which gases from the atmosphere are dissolved in the water, or why the solution of oxygen in a stream may be altered by transport mechanisms or by the natural growth of organic material, are more difficult to comprehend. Similarly, the concept of adsorption of some constituents on finely divided soils and/or the mechanics of the subsequent transport of these sediments remain foreign concepts to many. Yet, all these factors and many more provide the opportunity for extensive variance in the quality of the natural water system. [See *Environmental Geochemistry*.]

Table 3.2 provides a general description of how several selected characteristics of natural waters affect water quality. Table 3.3 lists the more prevalent natural inorganic constituents commonly dissolved in water and how they affect human use of the water.

Human preoccupation with problems of water quality, and with overall environmental quality, has increased substantially throughout the course of this century. During the first quarter of the twentieth century, emphasis was directed toward the need to minimize waterborne disease. By the end of the second quarter, emphasis began to focus in a significant way on issues of stream pollution with primary concerns directed toward reduction of organic and nutrient loadings and the impact of these loadings on stream biology. [See *Water Pollution and Toxicology*.]

An environmentally conscious public has continued to raise a host of issues. In rapid succession during the past decade concern has focused on several relatively new subjects. One of these is known as the trihalomethanes (THM) problem. THMs have been shown to be carcinogenic, and they can form as a reaction between chlorine and natural or manmade organic substances found in water. The paradox is that chlorine was successfully introduced in water treatment systems to meet the public health threat of waterborne pathogenic bacteria. Now the possible need to find other methods for achieving this disinfection or by better controlling the chlorination is prompted by a different public health issue.

Another major water quality issue of the past decade is the acid rain problem. Atmospheric discharge of wastes by certain industrial complexes is viewed as the primary cause of this problem. This issue has proven complex both scientifically and politically. Unanswered scientific concerns relate to the relative role played by humans and by nature in providing the offending constituents. Also, the explanation of why some water bodies are affected and other nearby lakes are not appears to require evaluation of a number of factors ranging from atmospheric transport mechanisms to the relative buffering capacity of both the tributary catchments and the receiving waters. [See *Air Pollution (Meteorology)*.]

TABLE 3.2

Selected Characteristics of Natural Water that Affect Water Quality[1]

Characteristic	*Principal cause*	*Significance*	*Remarks*
Hardness	Calcium and magnesium dissolved in water.	Calcium and magnesium combine with soap to form on insoluble precipitate (curd) and thus hamper the formation of a lather; also affects the suitability of water for use in the textile and paper industries and in steam boilers and water heaters.	USGS classification of hardness (mg/litre as $CaCO_3$): 0-60, soft; 61-120, moderately hard; 121-180, hard; 180, very hard.
pH (or hydrogenion activity)	Dissociation of water molecules and of acids and bases dissolved in water.	The pH of water is a measure of its reactive characteristics. Low values (below pH 4) indicate corrosive water that tends to dissolve metals and other substances that it contacts. High values (above pH 8.5) indicate alkaline water that on heating will tend to form scale. The pH significantly affects the treatment and use of water.	pH values: less than 7, acidic: 7, neutral; more than 7, bosic.

Specific electrical conductance	Substances that form ions when dissolved in water.	Most substances dissolved in water dissociate into ions that can conduct an electrical current. Consequently, specific electrical conductance is a valuable indicator of the amount of material dissolved in water; the larger the conductance. the more mineralized the water.	Conductance values indicate the electrical conductivity (in micromhos) of 1 cm^3 of water at a temperature of 25°C.
Total dissolved solids	Mineral substances dissolved in water	A measure of the total amount of minerals dissolved in water and is therefore a very useful parameter in the evaluation of water quality. Water containing less than 500 mg/litre is preferred for domestic use and for many industrial processes.	USGS classification of water based on dissolved solids (mg/litre): less than 1000, fresh; 1000-3000, slightly saline; 3000-10,000, moderately saline; 10,000-35,000, very saline; more than 35,000, briny.

[1]Heath, R. C. (1982). Basic Ground-water Hydrology, U.S. Geological Survey Water Supply Paper 2200, p. 65.

TABLE 3.3

Commonly Dissolved Inorganic Constituents in Natural Waters[1]

Substance	*Major natural sources*	*Effect on water use*	*Concentrations of significance (mg/litre)*[2]
Bicarbonate (HCO_3) and carbonate (CO_3)	Products of the solution of carbonate rocks, mainly limestone ($CaCO_3$) and dolomite ($CaMgCO_3$), water containing carbon dioxide.	Control the capacity of water in neutralize strong acids. Bicarbonates of calcium and magnesium decompose in steam boilers and water heaters to form scale and release corrosive carbon dioxine gas. In combination with calcium and magnesium, cause carbonate hardness.	150-200
Calcium (Ca) and magnesium (Mg)	Soils and rocks containing limestone, dolomite, and gypsum ($CaSO_4$). Small amounts from igneous and metamorphic rocks.	Principal cause of hardness and of boiler scale and deposits in hotwater hiaters.	25-50
Chloride (Cl)	In inland areas, primarily from seawater trapped in sediments at time of deposition; in coastal areas, from seawater in contact with freshwater in productive aquifers.	In large amounts, increases corrosivesness of water and, in combination with sodium, gives a salty taste.	250

Fluoride (F)	Both sedimentary and igneous rocks; not widespread in occurrence.	In certain concentrations, reduces tooth decay; at higher concentrations, causes mottling of tooth enamel.	0.7-1.2[3]
Iron (Fe) and manganese (Mn)	Iron present in most soils and rocks; manganese less widely distributed.	Stain laundry and are objectionable in food processing, dyeing, bleaching, ice manufacturing, brewing, and other industrial processes.	Fe>0.3, Mn>0.05
Sodium (Na)	Same as for chloride. In some sedimentary rocks, a few hundred milligrams per litre may occur in freshwater as a result of exchange of dissolved calcium and magnesium for sodium in the aquifer materials.	See chloride. In large concentrations may affect persons with cardiac difficulties, hypertension, and medical conditions. Depending on the concentrations of calcium and magnesium also present, sodium may be detrimental to irrigated crops.	60 (irrigation), 20-170 (health)[4]
Sulphate (SO_4)	Gypsum, pyrite (FeS), and other rocks containing sulphur (S) compounds.	In certain concentrations, gives water a bitter taste and at higher concentrations has a laxative effect. In combination with calcium, forms a hard calcium carbonate scale in steam boilers.	300-400 (taste), 600-1000 (laxative)

[1]Health, R. C. (1982). Basic Ground-water Hydrology, U.S. Geological Survey Water Supply paper 2200, p. 65.
[2]A range in concentration indicates potential level at which effect may become significant.
[3]From U.S. Public Health Service studies; optimum value dependent on water intake.
[4]Specific value dependent on individual diets and health.

A third major concern of the past decade is often described in the public media as the hazardous waste mess. Of major concern is the pollution of both surface waters and groundwaters with synthetic organic chemicals. Most of this pollution is initiated by leaching from point sources such as landfills, petroleum and other chemical storage tanks, pipelines, animal feedlots, and various industrial waste disposal sites. Other widespread hazardous waste pollution from nonpoint sources occurs as a result of the leaching of various agriculturally applied chemicals.

Table 3.4 provides a list of concentrations permitted in drinking water supplies by the U.S. Environmental Protection Agency as of the early 1980s. The reader should be alerted to the fact that these standards are undergoing continuous revision. The agency has also promulgated wastewater effluent standards that must be met.

TABLE 3.4

National Drinking Water Standards

Type of contaminant	*Contaminant*	*Maximum contaminant level (mg/litre unless marked otherwise)*
1	*2*	*3*
Primary regulations[1]		
Inorganic chemicals	Arsenic	0.05
	Barium	1.0
	Cadmium	0.01
	Chromium	0.05
	Lead	0.05
	Mercury	0.002
	Selenium	0.01
	Silver	0.05
	Fluoride	1.4-2.4 (varies with temperature
	Nitrate (as N)	10.0

1	*2*	*3*
Organic chemicals	Endrin	0.0002
	Lindane	0.0044
	Methoxychlor	0.1
	Toxiphene	0.005
	2, 4-D	0.1
	2, 4, 5-TP	0.01
	Total trihalomethanes[2]	0.10
Microbiological turbidity	Coliform bacteria	1/100 ml (monthly mean)
		1 TU (monthly ave.)
		5 TU (at state option)
Natural radioactivity	Gross alpha	15 pCi/litre
	Gross radium 226-228	5 pCi/litre
Manmade radioactivity	Gross beta	50 pCi/litre
	Tritium	20,000 pCi/litre
	Strontium-90	8 pCi/litre
Secondary regulations[3]		
	Chloride	250
	Colour	15 colour units
	Copper	1
	Corrosivity	noncorrosive
	Foaming agents	0.5
	Hydrogen sulphide	0.05
	Iron	0.3
	Manganese	0.05
	Odour	3 threshold odour number
	pH	6.5-8.5
	Sulphate	250
	TDS	500
	Zinc	5

[1]Federal Register, Vol. 41, No. 133, pp. 28402-28405, July 9, 1976.
[2]Federal Register, Vol. 44, No. 231, pp. 68624-68642, Nov. 29, 1979.
[3]Federal Register, Vol. 46, No. 140, pp. 42195-42202, Jan. 19, 1981.

Table 3.5 provides some comparative velocities for the flow of water in selected portions of the hydrologic cycle. These relative velocities serves to illustrate why groundwater pollution often goes undetected for a long period of time. They also explain the oft heard concern that pollution of groundwater is especially damaging because it takes so long for the pollution to disappear. The other side of the same coin argues that the very slow movement of groundwater localizes the pollution impact for a long time, and thus provides opportunity for possible containment and subsequent extraction of the polluted water. Indeed, the technology for achieving abatement of groundwater pollution is improving.

TABLE 3.5

Typical Range of Velocities at which Water Travels

Water classification	*Velocities (ft/day)**
Groundwater	0.01-10
Ice (glacier)	0.02-20
Surface water	50,000-500,000
Atmosphere	500,000-5,000,000

*1 fit/day=0-305 m/day.

D. Averages Versus Extremes

Humans have to adapt to the natural water resources system. This would not be a difficult task if management problems could be addressed in terms of average water conditions. Such is not the case. The management challenge comes in coping with periods of extreme excesses or extreme deficiencies. A gross measure of the magnitude of the task presented is best illustrated by citing some case examples.

National Weather Service analyses indicate that at a point in east-central Kansas the expected magnitude of the 6-hr rainfall that, on average, will be equalled or exceeded but once in 100 yr is 6 in. (152 mm). National Weather Service analyses also indicate that nature has the potential capability of delivering to this same location a total of 27 in. (686 mm) of precipitation in a 6-hr period. The range in observed streamflows can be even more dramatic. For example, in this same general location the Neosho River has, at a point where its drainage area totals 4905 square miles (12698 km^2) and the long-term average flow is 2510 cfs (71.5 m^3), experienced an instantaneous peak discharge of 410,000 cfs (11681 m^3). It has also experienced 5 successive months of zero flow and 60 successive months when the flow averaged less than 0.15 of the long-term mean.

The lesson contained in the foregoing statistics is quite clear. Human efforts at water management necessarily represent a calculated risk. Seldom, if ever, can society afford the magnitude of control works required to approach fail safe protection against either extreme flood or extreme drought.

I ANALYZING NATURE'S SYSTEM

The natural water system described above poses many challenging questions for the user society. Included are such issues as: What is the probability that a flood of a given peak rate of flow will occur? What is the average annual recharge to a specific well field? How much of the precipitation in a given storm will appear as direct storm runoff? How large a peak flow will that runoff produce at a target point downstream and when will that peak occur? If X amount of pollutant has been introduced at point A in a specified groundwater aquifer. how will the down-gradient plume be distributed over time? The potential list is unending. The paragraphs to follow will provide some insight into the techniques used to answer these questions.

A. Evaluating the Probability of Extreme Events

Critical questions of water management are necessarily associated with the occurrence of rare events. This implies a need to evaluate the probability of occurrence of an event of given magnitude so that risk considerations can be properly integrated into the decision process. Statistical appraisal of hydrologic events, however, faces several difficulties. Problems associated with analysis of streamflow data will serve to illustrate.

An initial problem relates to the adequacy of the available data base. In the United States a few of the major streams have adequate records extending over a 75- to 100-yr period. The bulk of records, however, relate to moderate-sized streams (drainage areas ranging from 100 to 5000 square miles [259 to 12944 km^2]) and seldom exceed 50 yr in length. The records for smaller catchments and for urbanized areas are for the most part less than 25 yr in length. In addition, many of these records must contend with significant statistical bias introduced by the changing culture and water use patterns introduced by society. In all cases the analyst must reconcile the question of whether or not the existing record represents a truly typical sample in the longterm world of natural phenomena. In recent years sophisticated modelling techniques, both deterministic and stochastic, have been introduced in an attempt to overcome possible deficiencies associated with short-term records.

The so-called persistence or lag inherently associated with the terrestial phase of the hydrologic cycle poses additional concerns. Conventional statistical analyses are based on the concept that each sample value represents an independent event. This is not true in the case of succeeding day streamflow. That is, the magnitude of the flow today is some function of yesterday's flow as tommorrow's flow will be a function of today's flow. A conventional assumption utilized in both high and low flow analysis is that the extreme experienced each year represents an independent event. Accordingly, the hydrologist performing frequency analyses normally works with a sample of annual extremes.

The end objective of frequency analysis is to define the probability p that a flow of given magnitude will be equaled or exceeded in any year. Hydrologists make widespread use of the average return interval T_R, which is the reciprocal of p, thus

$$p=1/T_R \quad \ldots(4)$$

A cumulative frequency curve can be calculated for the data contained in an annual series of extreme values. This requires that a decision be made on the plotting position assigned to each flow value. American practice, after many years of debate, has accepted the relationship

$$p=m/(n+1) \qquad ...(5)$$

where n is the number of events in the sample and m the rank order of the individual event. Thus, in flood analysis the highest event receives an m equal to unity. Conversely, in low flow analysis the m value of unity is assigned to the smallest value. Some recent studies indicate that the true probabilities of the upper end of the rank order are less than those produced by Eq. (5). (See *Probability*.]

Graphical plotting of the probabilities obtained from the foregoing equation requires that the cumulative frequency curve be eyeballed into place. For small return periods, that is, high probability of occurrence, this process is acceptable. At the higher return periods the process becomes most subjective, and there is much desire and theoretical persuasion to resort to the use of a mathematical distribution. One major readblock remains. No one has ever been able to prove conclusively that a specific probability density function is the correct choice for describing the distribution of a selected hydrologic variable.

Direct mathematical solution of the selected probability density function is rarely performed. Instead, hydrologists use the concept that most frequency relationships can be described by the following equation.

$$x=x+K\sigma_x \qquad ...(6)$$

where x is the random variable being analyzed, x the mean value of x, K a frequency factor dependent on the selected density function, and σ_x the standard deviation of x.

Values of K are readily available in tabular form for the most widely used distributions. For some distributions K is shown to be a function of the length of record as well as probability. For skewed functions K is also a function of the coefficient of skew.

The two most popular density functions currently being utilized in hydrologic frequency analyses are the log-Pearson Type III and the Gumbel Type I. The former has been formally adopted by interagency consensus within the U.S. government. The latter is the recommended choice of the major resources department in the United Kingdom.

For the log-Pearson Type III distribution the frequency factor K in Eq. (6) is a function of sample skew as well as probability. The log designation necessarily implies the substitution of log x for x in Eq. (6). The distribution is presumed to be skewed and is bounded on the left, but has the convenient mathematical characteristic of being equivalent to the standard log-normal distribution when the skew of the sample series is zero. However, there is little theoretical basis for the application of this distribution to hydrologic analysis. In fact, there are some dubious hydrologic characteristics associated with the use of this function. In the case of peak instantaneous flows the annual series will often contain values spanning three orders of magnitude, and in some cases four orders of magnitude. Under these conditions one often finds that the very low values in the series ordain the sign and magnitude of the skew. Thus, one is left with the paradox that the very small values in the sample series exert a disproportionate influence on the calculation of the magnitude of the rare floods utilized for design purposes.

For the Gumbel Type I distribution the frequency factor is theoretically a function of size of the sample as well as probability. However, conventional practice seldom incorporates any correction for the length of record, and the frequency factor is usually determined as

$$K=0.7797y-0.45 \qquad \ldots(7)$$

where

$$y=-\ln[-\ln(1-P)] \qquad \ldots(8)$$

This density function is based on statistical theories that argue that the distribution of extreme values, maximum or minimum, taken from n samples will approach a limiting form as the size of sample is increased. For the Type I version the distribution is given by

$$p=1-e^{-e^{-y}} \quad \ldots(9)$$

Hydrologists refer to this as the Gumbel function, named for its originator in hydrology usage. The rationale behind its use is that the maximum or minimum flow each year could be considered the extreme values in a sample of 365 entries.

There are numerous other frequency or probability issues encountered in day-to-day water management. Many of the questions posed relate to problems of conditional frequency. Direct comparison of rainfall and runoff frequencies will not, for example, be useful unless the watershed is a seamless, small, clear sheet of impermeable material. This is so because individual events in the runoff series are not only dependent on the current precipitation, but are also influenced by the recent antecedent precipitation and evapotranspiration history. Similarly, joint frequency issues arise in the design of such works as urban storm drains. In this case the frequency of the storm discharge from the urban drain will normally differ from the frequency of the stage in the receiving waters. Accordingly, the risk evaluation of project performance necessarily involves resolution of a joint frequency problem.

A fundamental conclusion is that risk assessment in water management is a must, and there is need to improve performance in this area of inquiry.

B. Selected Deterministic Event Models

Determinstic methods used in analyzing the natural water system are based on established physical principles or on conceptual relationships presumed to describe the pertinent physical conditions. Inasmuch as fluid flow is a significant consideration, many of the analytical tools are derived from fundamental principles of fluid motion. In many cases the basic mechanics of flow have to be supplemented by perceptive choices of both empirical coefficients and empirical relationships. The major consideration forcing the use of empirical tools relates to the analyst's inability to control the boundary geometry. This is to say that no two storms, catchment areas, channels, aquifers, or antecedent moisture histories are exactly alike.

Deterministic solutions utilize mathematical models and can be cast into the two general categories of event models and continuous models. Event models allow the solution of specific problems within the overall system, given presumed or existing boundary conditions. They are often thought of as designer models. Continuous models altempt to calculate system response for long periods of time, and for whatever combination of boundary conditions evolve during that period of time. Continuous models, therefore, are useful to the designer in evaluating the operational performance of alternative management actions. They necessarily serve the needs of the operational water manager.

Space constraints prohibit an exhaustive discussion of all the event type problems one can encounter, but a brief discussion of selected problems and underlying solution concepts is important before directing attention to the continuous type model.

For very small catchments an assumption of uniform catchmentwide precipitation is acceptable, and for selected design problems an assumption of uniform rainfall over areas of 100 square miles (160 km^3) or more is not uncommon. However, in attempting to reconstitute the hydrologic performance of a specific storm or fulfilling the operational demands of day-to-day forecasting, the analyst is confronted by nonuniform precipitation patterns. This problem is compounted by the fact that available precipitation stations are not unformly distributed. An initial problem, therefore, is the determination of a weighted average precipitation for the catchment or segments thereof.

The most widely used methodology for solving this problem is the Theissen polygon. The procedure requires (a) plotting to scale the location of the gauging locations within and surrounding the catchment, (b) drawing connecting lines between stations, (c) constructing perpendicular bisectors to each connecting line so that the bisectors form a polygon around each station, and (d) determining the area of that portion of each polygon that is contained within the catchment and the percentage of total catchment area it represents. The precipitation at each gauge is assumed to apply to the entire polygon surrounding the station. The weighted average precipitation for the total catchment, or segment thereof, is the sum of the products of the individual gauge precipitation and percentage of the area assigned to the gauge.

The process just described necessarily presents a significant opportunity for error. It is based on an assumption of linear variation between stations and the assignment of a constant segment of area to each station. In actuality neither assumption may be accurate; especially if gauge density is small or if orographic influences are significant. Moreover, if gauge density is small and storm patterns are of a convective nature and do not extend from gauge to gauge, the effective source area influenced by a gauge is not constant storm to storm. New advances in remote sensing technology, primarily the use of radar with digital mapping of imagery, are directed toward overcoming many of these problems.

A major and repetitive problem encountered in the analysis of the natural water system is that of determining how much of the water in a given precipitation event will appear as direct storm runoff. The general area of inquiry is often referred to as the process of defining the rainfall-runoff relationship. This relationship is a function of the physical characteristics of the catchment, antecedent climatic characteristics, and storm characteristics. The controlling mechanism is infiltration, which is the passage of water through the soil surface. Water that enters the soil will flow by gravity through noncapillary channels permeating all root zones. However, capillary forces divert some of this water into available capillary pores. This serves to increase the resistance to flow and, in turn, leads to a decreasing rate of infiltration with time. The classic work of Horton produced the following expression for the rate of infiltration.

$$f_p = f_c + (f_0 - f_c)e^{-kt} \qquad \ldots(10)$$

where f_p is the infiltration capacity expressed as a rate (length/time f_c the constant infiltration rate that can be sustained after the soil zone has become saturated, f_0 the infiltration capacity at the beginning of the storm, e the napierian base, k an empirical coefficient, and t the time since start of the storm. Theoretical application of this equation presumes the latter to be applicable only when the precipitation intensity exceeds the infiltration capacity; otherwise the infiltration capacity is presumed equal to precipitation intensity.

Research has shown that the infiltration rate is dependent on many factors, such as soil characteristics, vegetal and man-made cover, surface slope, and the prevailing moisture content of the soil. Since all of these factors may vary significantly across a catchment and some exhibit a pronounced temporal variance, any general solution would necessarily be very complex. Fortunately, for any given catchment many of these parameters may be taken as constants. Accordingly, the predominant factors in defining the rainfall-runoff characteristics of a specific catchment become the antecedent climatic characteristics and the individual storm characteristics.

The antecedent climatic characteristics establish the soil moisture condition at the start of the storm. In the context of the Horton equation they define the value of f_0. This moisture condition is determined by the combined history of antecedent precipitation and antecedent evapotranspiration. In theory, the antecedent history should extend over the infinite past. Practice has shown that the initial choice of a soil moisture value is not critical for continuous moisture accounting models after several weeks of operation. For event design models the moisture content is often arbitrarily incorporated into the design criteria.

For the needs of operational hydrology absolute values of the soil moisture are not necessary. What is needed is an index of soil moisture that can be utilized with other pertinent parameters in a multivariate correlation relationship. In the absence of day-to-day calculation of evapotranspiration the antecedent precipitation is often decremented logarithmically to provide a continuous value of a moisture index. Thus

$$I_t = I_0 k^t \qquad \text{...(11)}$$

where I_t equals the antecedent moisture index at time t and I_0 equals the initial antecedent moisture index. Inasmuch as evapotranspiration has a seasonal variation, the value of k needs to vary with the calendar or some other calendar parameter needs to be included in the multivariate relationship.

Pertinent storm characteristics are twofold: both the amount of rainfall and the effective duration of rainfall are needed. That is, a 3-in. rainfall received in a 1-hr period produces a different response than a 3-in. rainfall roughly distributed throughout a 24-hr period.

In summary, the rainfall-runoff relationship for a specific catchment is controlled by the amount and duration of the storm and the antecedent moisture condition at the start of the storm. The latter is controlled by the combined recent history of moisture applied (precipitation) and moisture withdrawn (evapotranspiration) from the soil zone.

Once the magnitude of runoff volume has been determined, questions arise concerning the time distribution of that runoff at selected downstream targets. This general area of inquiry is known as hydrograph theory.

Consider an imprevious catchment of unit width and length L placed on slope S. Apply water to this watershed from an overhead sprinkler at a uniform rate i. Measure the rate of flow of the water as it exits the catchment. If this measured flow, (the catchment discharge), is plotted against time, an ogee type curve will result. That is, initially the discharge will increase slowly as volume develops on the surface and surface resistance to flow is overcome. Subsequently, the rate of flow will increase much more rapidly and then the rate of increase will slow perceptibly as the exit discharge approaches equilibrium with the inflow. The equilibrium flow would necessarily be i(L) (1) and would have units of L^3/t. Until equilibrium is reached, the inflow to the catchment is greater than the outflow from the catchment. The integrated volume of this difference represents the storage accumulated on the catchment. Once equilibrium conditions are attained, the storage remains a constant. If at a later time the inflow is discontinued, the outflow will begin to diminish—rapidly at first and then slower as it approaches zero. The plot of outflow with time is illustrated by the solid lines in Fig 3.1A. This plot of discharge versus time is called a hydrograph. The increasing and decreasing portions of it are termed the rising and recession limbs.

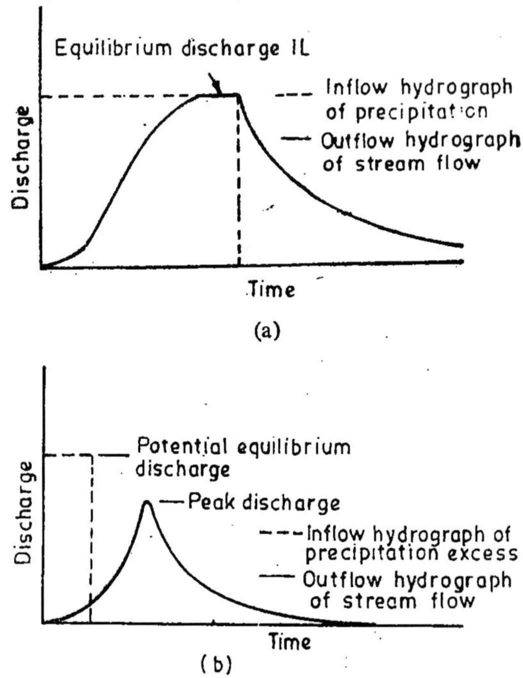

Fig. 3.2: Example hydrographs. (a) Equilibrium hydrograph from impervious elemental strip catchment. (b) Typical hydrograph from normal catchment

One can intuitively visualize some changes to be expected in the outflow hydrograph as certain of the foregoing parameters are changed. For example, if the size of the catchment were increased by increasing the length L, the magnitude of the equilibrium discharge would increase and the time to equilibrium would also increase. Increasing the slope would serve to shorten the time to peak and decrease the amount of basin storage at equilibrium. Alternatively, inserting a small dam in the middle of the catchment would delay the time to peak and alter the shape of both the rising and recession limbs.

Some of the rudimentary concepts illustrated by this elemental catchment hydrograph can be extended to evaluation of hydrographs on larger catchments. The significant differences are these. First, natural catchments are not impervious. Thus, an equilibrium discharge if reached would not be the product of iA, where A is catchment area, but rather $i_e A$ where i_e is the rate of rainfall excess. Second, except on very small catchments, the generation of rainfall excess does not continue until equilibrium conditions are reached. Instead, the catchment is pulsed with a burst of rain and the subsequent exit hydrograph peaks at some later time. This concept is illustrated by the sketch in Fig 3.1B. Note that the time and discharge scales in Fig 3.1A and 3.1B are necessarily different. The exit peak flow is less than the peak of the input pulse. This reduction in flow rate results from the combined effects of catchment lag and basin storage.

Third, the storage characteristics in the catchment are far more complex than the simple surface detention of the elemental catchment. They include, in addition to storage on the plane surface, storage in natural and artificial impoundments, storage in overflowing floodplains, storage within the channels, and storage in discharging groundwater aquifers. Accordingly, the recession or falling side of catchment hydrographs reflect varying contributions from these several sources. Nonetheless, the general concepts illustrated by the elemental catchment hydrograph are honoured on larger watersheds. For example, between two otherwise similar watersheds the one having more storage will show a longer time to peak and a flatter peak.

Finally, watershed response can be altered by the activities of humans. Fig. 3.2 illustrates possible change associated with urbanization. In this case the insertion of extensive impervious areas will increase runoff volume but decrease groundwater recharge and thereby alter recession characteristics. Also, in the absence of specific mitigation efforts, the magnitude of peak flow can be expected to increase and the time to peak decrease.

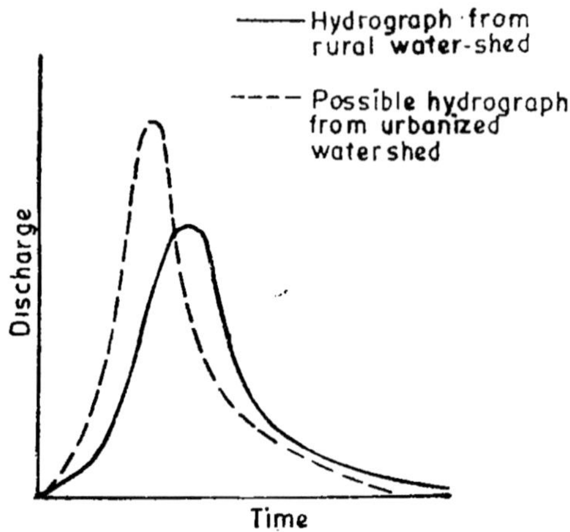

Fig. 3.3: Hydrograph impacted by human-induced cultural change

Hydrographs of concern are generally categorized as either overland flow hydrographs or streamflow hydrographs. The former corresponds closely to the physical conceptualization associated with our elemental catchment, that is, where detention on the plane surface is the only storage medium, Streamflow hydrographs, in contrast, are impacted in a major way by the lag time through the catchment and by channel and flood-plain storage characteristics. The overland flow problem can be solved by direct application of flow equations. Empirical concepts are the more common rule in the analysis of streamflow hydrographs.

The basic equation of motion for fluid flow can be written

$$S_f = S_0 - \frac{\partial y}{\partial x} - \frac{v \partial v}{\partial x} - \frac{1}{g}\frac{\partial v}{\partial t} \qquad \text{...(12)}$$

where S_f is the slope of the energy gradient, S_0 channel slope, y the depth of flow, υ the velocity, x the distance along the channel, t the time, and g the gravitational acceleration. When $S_f = S_0$ the flow is classified as steady uniform. Addition of the second and third terms on the right side of Eq. (12) produces the equation for steady nonuniform flow. Addition of the last term enables consideration of unsteady nonuniform flow.

The basic equation of continuity for unsteady incompressible fluid flow can be written

$$\frac{\partial Q}{\partial x} + \frac{B\,\partial y}{\partial t} = 0 \qquad \text{...(13)}$$

where Q is the total discharge and B the surface width of the flowstream. For a channel of unit width, Eq. (13) may be rewritten

$$\frac{\partial q}{\partial x} + \frac{\partial y}{\partial t} = 0 \qquad \text{...(14)}$$

The overland flow problem can be viewed as similar to the elemental catchment with i_e (the rate of rainfall excess) taken as lateral inflow applied uniformly along the x axis. In this case the continuity equation per unit width becomes

$$\frac{\partial q}{\partial x} + \frac{\partial y}{\partial t} = i_e \qquad \text{...(15)}$$

Overland flow is encountered at the head end of natural catchments before the flow becomes channelized and in manmade parking lots, taxiways, etc. In these cases the S_0 term is usually much larger than the other terms on the right side of Eq. (12). Accordingly, an assumption that $S_f = S_0$ is not inappropriate. For this condition fluid mechanics provides the following expression for q:

$$q = ky^m \qquad \text{...(16)}$$

where the values of the coefficient k and the exponent m are dependent on whether the flow is laminar or turbulent.

Simultaneous solution of these several expressions following what is known as the kinematic wave theory produces the following equations for time to equilibrium

$$t_e = 0.94\,\frac{n^{0.6}L^{0.6}}{S^{0.3}(i_e)^{0.4}} \qquad \text{(turbulent)} \qquad \text{...(17)}$$

$$t_e = 9.3 \frac{v^{0.333} L^{0.333}}{S^{0.333} (i_e)^{0.667}} \quad \text{(laminar)} \qquad ...(18)$$

where the constants presume the use of english units and provide t_e in minutes with L in feet. The symbol n in the turbulent equation is the manning roughness coefficient and v in the laminar equation represents the kinematic viscosity. Experimental evidence has shown that the overland flow hydrograph for both turbulent and laminar flow can be collapsed to a dimensionless plot of q/q_e versus t/t_e. Inasmuch as q_e and t_e can be determined by the above equations, solution of the overland flow hydrograph represents a nice tidy package.

The major reason for citing the foregoing solution is to be able to contrast it with techniques used on the larger channelized catchments. Note that Eqs. (17) and (18) clearly indicate a nonlinearity in the time response as the rate of rainfall excess is varied, One should also note that some investigators have used the kinematic wave and other dynamic solutions in addressing questions of hydrograph development on larger channelized catchments. These attempts indicate the system response to be nonlinear. The most common technique used in hydrograph theory is based on an assumption of linearity. The concept is known as unit hydrograph theory or unitgraph theory. A unit hydrograph by definition is the hydrograph of direct storm runoff produced when a unit depth of rainfall excess is generated in a t period pulse and for a specified areal pattern. The conceptual model argues that a storm having the same areal pattern and of equal pulse duration t, but which generates three units of rainfall excess, will produce a hydrograph whose ordinates are three times those of the unit hydrograph.

For reasons implied above numerous mathematical objections to the unitgraph concept have been raised. Moreover, simple logic argues against the validity of the linearity assumption. For example, consider two hydrographs, one of which contains a unit volume and the other five units of volume, both generated on the same watershed in equivalent length pulses. Examine only the recession side of the hydrograph. The point of contraflexure on the recession side is presumed to be the point at which no more new inflow is being added to the flow stream at the hydrograph site. Thus, all flow following contra-flexture represents drainage from storage. The second case has far more water in storage than the first and the time to release it should be greater in the second than in the first. Yet, unit-graph theory would produce the same base length for the two hydrographs.

Regardless of the criticisms implied above, the fact remains that unitgraph theory has been widely and relatively successfully utilized for almost a half of century. It is true that unitgraph peaks derived from small events are often smaller than those obtained from larger storms. The foregoing observation plus the implications associated with the mathematical concepts of fluid flow in channels have led some hydrologists to increase peaks and decrease time to peaks by guestimated amounts when performing design calculations associated with extreme floods. Such adjustments may or may not be prudent. Sometimes when streams overflow their main channels the shortening of travel time otherwise anticipated with increasing discharge in a given channel may not occur. In the final analysis the success of the unitgraph concept probably relates to the lumpiness inherent in the overall problem. The averaging process of non-uniform precipitation and nonuniform antecedent moisture plus the modulating effect of irregular flow sections makes isolation of nonlinear impacts among different storms difficult to quantify. In short, increased accuracy is not assured at this point in time by the application of complex nonlinear models. Accordingly, the ease of utilization associated with the linear assumption has much to do with the method's acceptance.

In practice unitgraphs are developed from observed records on gauged streams. Regional analysis of observed unitgraphs is used to provide empirical relationships for unitgraph derivation on ungauged areas. Standard techniques require predictive equations for time to peak, magnitude of peak, and length of base. Standard dimensionless hydrographs provide guidance in drawing the rising and failling limbs by trial and error until the unit volume requirement is met. In application the underlying assumption of linearity is extended to the concept of superposition in the analysis of complex storms. Thus, a 2-hr unitgraph may be used to estimate the hydrograph produced by a nonuniform 4-hr storm in which the first 2-hr period generates 0.5. in. of rainfall excess and the second 2-hr period produces 3.0 in. of rainfall excess. Solution requires multiplying the unitgraph ordinates by 0.5 and by 3.0, but lagging the 3.0 products 2 hrs behind the 0.5 products. The sum of the two columns will then represent the total storm runoff.

There is often need to address the issue of how hydrographs are altered as they move along a stream, the process is referred to as hydrograph routing or hydrograph translation.

Prior to the availability of digital computing capability this problem was approached almost exclusively by what is known as hydrologic routing. This technique is empirical and is still widely used. The procedure is based on continuity of mass so that

$$I-O=\frac{\Delta S}{\Delta t} \quad \ldots(19)$$

where I is the mean inflow to a given segment of the stream, O the outflow from the segment, and ΔS the change in water volume within the segment during the time element Δt. Difficulty arises because (a) the flow is unsteady and (b) the outflow at the end of the reach lags the inflow at the head of the reach. Thus, there is no unique relationship between stage at either end of the reach and storage volume within the reach. The conventional hydrologic approach to this problem is of the form

$$S=K[XI+(1-X)O] \quad \ldots(20)$$

where S equals storage in the reach, K a constant of proportionality, and X a dimensionless weighting factor. Examination of the equation indicates K represents a volume divided by a volume rate and thus has the units of time. It is often assumed to be equal to the travel time of a flood wave moving through the reach. The value of X theoretically varies from 0 to 0.5, with most common values being between 0.2 and 0.3. One should note that an X value of zero makes storage solelv a function of the outflow which equates to a level reservoir condition.

The advent of the digital computer has enabled the introduction of new hydraulic techniques as contrasted to hydrologic solutions. These hydraulic approaches are based on fundamental fluid mechanics, namely, the conservation of mass (volume in an incompressible fluid) and the conservation of momentum. Many solutions using the complete equations have been presented, and numerical methods based on both explicit and implicit schemes have been used. Strict adaptations of the complete equations require evaluation of the inertial response to all pertinent forces. Some of the more popular solutions include simplifying assumptions. The previously mentioned kinematic solutions consider only gravity and friction forces, that is, the energy slope is set equivalent to the gravitational slope. Other schemes omit only the inertial response. Critical questions concerning implementation of hydraulic routing schemes relate to costs, for example, field surveys required to define channel cross-sections, stability of the numerical schemes, and the assignment of hydraulic parameters such as channel roughness.

Water that does not become direct runoff infiltrates the soil. The subsurface movement of water presents a number of problems that demand quantitative appraisal. Three selected subjects will serve to illustrate. These are

(a) flow in the saturated zone,

(b) the downward flow of water in the unsaturated zone, and

(c) hydrologic response to pumped withdrawals.

The subsurface can be divided into saturated and unsaturated zones. Only water in the saturated zone supplies springs, streams, and wells and is the water nominally classified as groundwater. The unsaturated zone is usually subdivided into three parts called the soil zone, intermediate unsaturated zone and the capillary fringe. The first extends downward from the land surface about 2 to 3 m depending on the local geology.

Water in this zone is charcterized as soil moisture. The capillary fringe results from the attraction between water and solids and extends upward a few centimeters from the saturated zone. The intermediate unsaturated zone contains water that is percolating from the soil zone to the saturated zone. Water in the unsaturated zone is under a negative pressure, whereas the water table at the top of the saturated zone is equal to atmospheric pressure.

Flow in the saturated groundwater is normally presumed to follow Darcy's law for flow through a porous media. Thus, velocity is stated to be proportional to the slope of the hydraulic gradient. The discharge is the product of velocity and area, but the effective area is the cross-sectional area multiplied by the porosity of the medium. Thus,

$$Q = kpAs \qquad \text{...(21)}$$

where Q is the discharge, k the Darcy coefficient of proportionality, A the normal cross-sectional area, p the porosity, and s the hydraulic gradient. This equation is normally written

$$Q = KAs \qquad \text{...(22)}$$

where K, the product of k and p, is called the coefficient of permeability or the hydraulic conductivity. Often A is projected as a rectangle of saturated depth Y and saturated breadth B. In this case Eq. (22) is rewritten

$$Q = TBs \qquad \text{...(23)}$$

where T is the product KY and is known as the transmissibility or transmissivity of the aquifer.

Flow through the unsaturated zone is termed percolation and is extremely important because that is the way recharge of the saturated groundwater occurs. This flow is a function of both gravitational and capillary forces. An assumption of steady-state flow allows one to use a modified form of the Darcy relationship. The modified form requires a hydraulic conductivity that is less than that obtained for the same material in a fully saturated zone. It also requires a slope term that becomes the sum of the capillary gradient and the gravitational gradient. An assumption of steady flow however, is not very realistic where natural recharge is involved. It has more application in human created environments where one may be trying to track the seepage from a brine pond or leachate from a waste fill. Flow in the unsaturated zone can be most sensitive to nonuniform subsurface materials. Thus, stratification which is very much a part of natural geologic evolution serves to complicate analysis in this zone, and very precise subsurface information may be required to track flow in the nonsaturated zone. Overall, analysis of unsaturated flow can be characterized as most complex. It is the subject of much current research.

Any analysis of subsurface gravitational flow requires knowledge of the hydraulic properties of the saturated aquifer. Prior reference was made to the hydraulic conductivity and/or transmissivity. The other critical property is the storage coefficient, which is defined as the volume of water that an aquifer releases per unit of surface area per unit drop in head.

When a well is inserted in an aquifer and pumped, a cone of depression develops in the water table surrounding the will. This occurs because the velocity must increase as flow approaches the well and the circumferential area normal to the flow is decreased. The amount by which the hydraulic level in this depressional cone is lowered is called the drawdawn.

Nineteenth century solutions assumed that the cone of depression reached an equilibrium condition. That does not happen in the absence of extensive recharge or long pumping periods. More recently solutions based on a heat flow analogy have been found applicable to unsteady conditions. The basic equation for calcuating drawdown is

$$Z = \frac{QW(u)}{4\pi T} \qquad \ldots(24)$$

where Z is the drawdown, Q the pumping rate T the transmissibility, and W(u) given by

$$W(u) = 0.5772 - \ln u + u - \frac{u^2}{2 \times 2!} + \frac{u^3}{3 \times 3!} - \frac{u^4}{4 \times 4!} + \cdots$$

and

$$u = C\frac{r^2 S_c}{4tT} \qquad ...(25)$$

where S_c is the aquifer storage coefficient, r the radial distance from the well, t the duration of pumping, and C a constant, the magnitude of which is dependent on the flow units selected.

The foregoing equations assume T remains constant during the pumping period that the water withdrawn is obtained entirely from storage, that the well penetrates the entire thickness of the aquifer, and that the aquifer is unbounded. Given the aquifer, characteristics, Eqs. (24) and 25 can be used to predict aquifer response to a proposed or existing pumping pattern. The concept of supperposition is used in accumulating the effects of two or more wells. Care must be taken to correct for the effect of recharge where it is meaningful and to correct for the effect of boundary conditions, for example, maintenance of a constant head due to leakage from a surface source or increased dewatering imposed by the effect of an impermeable boundary. Equations (24) and (25) can also be used to determine aquifer characteristics. In this case a well is test pumped and data on drawdown and time are obtained at one or more observation wells. These data can then be utilized in conjunction with graphical techniques and Eqs. (24) and (25) to solve for the aquifer properties.

Prior to the development of both analog and digital computers the analyses of the flow patterns in a large aquifer and or large well field were onerous. In the late 1950s and early 1960s groundwater problems were studied with electric analog computers. In this computer, current is analogous to flow, voltage is analogous to potential head, and permeability or conductivity is established as the reciprocal of resistance. The specific characteristics of an aquifer, including pumps, recharge sources, etc., are reproduced by a network of resistors and capacitors. By inputing or withdrawing current at vorious points in the circuit, the model can be used to analyze wellfield and aquifer performance. In general, the electric analog approach solves the basic differential equation of groundwater flow from which Eqs. (24) and (25) are developed.

With the advent of the digital computer and numerical methods of solution, the costly and one aquifer electric analog was displaced by new digital models. Several hundred digital models for solving groundwater flow problems have been documented. For the most part they utilize finite difference methodology or related numerical technique. They necessarily require the analyst to depict critical boundary conditions in mathematical form. The latter include such items as inflow recharge, impermeable strata, and point and line sources. Output narmally includes the time series of groundwater levels at the node points within the modeled area. Major errors associated with these models derive from poor assumptions regarding boundary conditions and/or aquifer characteristics. Criticism is often voiced that many of these models presume or seek an accuracy not contained in the input data and assumptions. Two-dimensional-models are far more popular than three-dimensional efforts due to cost and complexity considerations. The greatest current demand for these models involves their coupling with contaminant transport concepts to appraise groundwater pollution issues. [See *Numerical Analysis.*]

C. Continuous Models

Continuous models have undergone substantial evolution within the past two decades, again in response to the development and refinement of digital computer equipment. These models have found acceptance in the extension of short-term flow records and in the analysis of the hydrologic performance of ungauged areas. Also, they are widely used in the comparative analysis of alternative water management schemes.

Note that the squares in the chart represent flow processes, the rectangles represent storage zones, and the triangles represent input and output. The underlying principle of these models is that of hydrologic mass balance. Climatic input data (also any human-induced imports or exports) are fed into the model for a specified time interval Δt. Before input for the next time increment is accepted, the model via appropriate logarithms will determine how much water is added to or extracted from each of the storage zones. Equation (19) is used to adjust the water content of each zone during each time increment.

The simulation logarithms are based on the event concepts discussed previously. For example, if one uses a uniform duration Δt of rainfall input, then the storm characteristics are characterized by the amount of precipitation applied in Δt. The primary remaining variable influencing the split between initiation of surface flow and infiltration becomes the soil moisture. Accordingly, an algorithm that relates this split to a soil moisture index at the start of Δt is utilized. Note that in these simulation techniques an absolute value of soil moisture is not required; the necessary ingredient is a relative index value.

The degree of sophistication incorporated is dependent on both the utilization being made and the adequacy of the input data. For example, if the objective is appraisal of the adequacy of a given water source, daily calculation of streamflow with printed monthly outputs may suffice, If, instead, flood mitigation studies are the subject of concern, hourly outputs might be required. In the latter case a subroutine involving some from of hydrograph theory and an hourly printout of streamflow at selected target points in the catchment would be incorporated.

When catchmentwide average values of precipitation, soil moisture storage, surface slopes, and cover are used the simulator is characterized as a lumped model. The alternative of dividing the catchment into subsections to that variations in these and other parameters will better represent catchment characteristics is known as a subdivided model. In theory, a subdivided model will give a more accurate solution than a lumped model. This may not be true. Although there may be ample guidance for programming some variance in catchment characteristics, the available climatic input may not justify subdivided analysis. In such cases the overall performance of a subdivided model may be no better than that obtained with a lumped approach. If so the added operating costs of the subdivided model will seldom be justified.

Continuous simulators require calibration runs against observed records obtained on the catchment being studied or on a geologically and locationally similar catchment. These calibration efforts involve the trial and error adjustment of various parameters utilized in the simulator. This process is often criticized on grounds that it does not ensure a unique solution. Thus, it may be argued that extended operational studies based on short-term calibration may lead to a misleading appraisal of longterm performance.

In a rigorous mathematical sense this type of criticism is valid. Nonetheless, experience with these models has been most encouraging, and there are certain techniques that mitigate the possibility of poor performance. For example, calibration should not be limited to comparison with one outflow hydrograph if there is opportunity to compare simulation output with two or more gages. In some cases investigators have simulated not only catchment outflow, but have also fed the indicated groundwater recharge from the continuous simulators as inflow to a finite difference groundwater model and matched observed groundwater levels as well as stream outflows.

For the uninitiated, the number of parameters in a continuous model triggers concern over possible performance. For example, the Stanford Watershed Model, often viewed as the pioneering effort in this field, utilizes 15 parameters. However, over half of these are-quantified from maps, observed climatic and flow data, and direct field examination. As such they are not subject to change by calibration. The potential range of several others is small and their values can be closely related to geologic characteristics. Thus, calibration really involves the reconciliation of about four parameters. This still leaves opportunity for multiple solutions, but repeated testing of a given simulator on a variety of catchments has shown that these key parameters also possess significant correlation with identifiable physical characteristics.

Numerous papers have advocated simulator calibration through use of an objective function. A common suggestion is to minimize the sum of squared differences between observed and computed daily flows. Care must be exercised in utilizing such procedures. An error in the input data for a major event will produce not only a substantial error in the associated peak discharge, but due to the persistence in the system, a continuing substantial error for a number of subsequent days. A model that replicates the daily flow duration curve and exhibits no bias in the time to peak or magnitude of peak flows can be expected to provide useful perspective.

D. Stochastic Models

The deterministic processes described in the two prior sections seek to simulate a specific event or a continuous stream of specific events. By way of contrast, stochastic hydrologic models are not event related. Instead, they seek to produce and examine alternative time series traces of a hydrologic parameter subject to the constraint that each trace is presumed to have an equal opportunity of occurrence. [See *Stochastic Processes*.]

The motivation for development of stochastic models relates to earlier comments about the short-term nature of available hydrologic data and difficulties encountered in evaluating the probable occurrence of extreme values. For example, in water supply analysis determination of the size reservoir needed to supply a given flow throughout the drought of record may well beg the issue. The unanswered question is how extreme was the drought of record.

A stochastic modeler will aproach this problem with several basic assumptions. First, the available x years of record do not represent a unique time trace, that is, the next x year period will generate a different trace. Second, there is every reason to believe that either x year trace has an equal chance of occurrence. Third, although the range and sequence of the traces will differ, one can assume that both traces, and all other x year traces, will display consistent statistical behaviour. Thus, even a relatively short x year sample period is sufficient to determine with reasonable accuracy such basic population parameters as the mean and standard deviation or, when two or more variables are involved, such parameters as regression coefficients and standard errors of estimate.

Returning to the water supply example, the stochastic approach would require (1) determining the appropriate statistical parameters from the available record; (2) selecting a reasonable economic life for the proposed reservoir; (3) generating a large number of flow traces, each equal to the specified economic life of the project; (4) finding the storage necessary to provide the required flow for each such trace; and (5) making a probability analysis of the storage values so obtained. The conceptual basis is easy to accept. Critical points in the process involve steps (1) and (3): the selection of the critical statistical parameters and the mechanism for generating the alternative traces. A typical generator incorporates autoregressive techniques that attempt to capture seasonal serial correlations. For example, the basic equation for a monthly generation of flow is

$$q_{i,j} = q_j + \rho_j(\sigma_j/\sigma_{j-1})(q_{i,j-1} - q_{j-1}) + t_{i,j}\sigma_j\sqrt{1-\rho_j^2} \qquad \ldots(26)$$

where the subscript i is the serial designation of years ranging from 1 to n, and the subscript j refers to the month of the year ranging from 1 to 12. The terms σ_j and σ_{j-1} represent the standard deviations of the recorded monthly values of q for their respective months. The serial correlation coefficient ρ_j relates q_j to q_{j-1}, and the product $\rho_j(\sigma_j/\sigma_{j-1})$ represents the regression coefficient for the q_j and q_{j-1} relationship. The t value in the final term represents a random number based on whatever distribution is presumed to apply.

Equation (26) basically (states that the value of $q_{i,j}$ will be the historic mean value of q_j adjusted for the serial influence (presumed to represent the hydrologic persistence in the system) introduced by q_{j-1} and further adjusted by the random impact of the weather in month j. A key decision is the selection of the probability function needed to characterize the distribution of t. Moreover, it is conceivable that the distribution of t is not the same at both high and low values of discharge. A widely used assumption is that of a normal distribution with zero mean and unit variance around the mean value of q_j for a given value of q_{j-1}. Experience has shown that utilization of a log-normal distribution is often preferable. Many papers have been prepared showing that the distribution that best preserves the underlying characteristics varies from stream to stream. Thus stochastic methods are also subject to calibration questions. Unless the random variate t is selected from a proper distribution the model will not produce a proper trace even if it has preserved such parameters as the mean and variance.

Basic to the stochastic concept is the inherent assumption that the process is stationary. If, indeed, the statistical properties do vary with time, the approach should not be utilized without first adjusting the record to account for changing physical conditions. An example would be the case of an urbanizing watershed where the installation of impervious surfaces is changing the basic yield characteristics.

Experience has also shown that utlization of Eq. (26) cannot precede oblivious to the need to check other than monthly results if pertinent underlying characteristics of the initial record are to be preserved. If, for example, the serial correlation of annual flows is less than 0.1 direct application of the equation has been found acceptable. However, if there is significant serial correlation in annual flows, this correlation will not be assured by Eq. (26). In such cases there is a growing body of evidence that generation of annual flows with a disaggradation to monthly values will improve results. Similarly, attempts at generating subdivided flows and/or areal rainfall patterns need to examine the preservation of imerstation correlations.

Other techniques have been advanced to improve present capability in the hydrologic application of stochastic concepts. Many critics have suggested that the concept of preserving the high-frequency characteristics of short-term records in no way ensures additional distribution of the low-frequency events. Accordingly, many new efforts are attempting to address this question. Similarly, most success has been achieved with monthly or annual models. Attempts to model time intervals of less than a month have encountered some substantial obstacles. A major hurdle is the proper definition of the random variate distribution. Some writers have suggested that stochastic models that reflect short duration events will have to incorporate some deterministic elements.

E. Water Quality Models

Society has become increasingly preoccupied with the need for water pollution control. This need has generated much interest in the development of water quality or contaminant transport models. These models can be described in a number of different ways. From a water viewpoint one can visualize a surface water model or a groundwater model. These can be further classified as steady-state or unsteady-state water flow conditions. Similarly, one can have steady or unsteady contaminant streams. From a pollutant or constituent viewpoint one can talk about point source and nonpoint source models. Also, models can be classified as conservative or nonconservative depending on whether the mass of the specified constituent is altered by chemical or biological reactions.

The partial differential equation defining the concentration of any constituent in a one-dimensional flowstream is

$$\frac{\partial C}{\partial t} = \frac{\partial \left(KA \frac{\partial C}{\partial X} - VAC \right)}{A\,\partial X} \pm \sum_{n} S_n(V, X, t) \qquad \ldots(27)$$

where C is the constituent being traced, t the time, X the longitudinal distance along the flowstream, K the coefficient of dispersion, A the cross-sectional area of flow, V the velocity, and S any source or sink additions or subtractions of C. The equation states that at a specified point, the change in concentration with time is a function of the dispersion and advection properties of the flow stream plus sources of C and minus sinks of C. The latter items include any reactions or additional loadings that change the value of C for the point in question.

For a wide range of quality concerns critical concentrations occur during low water periods when streamflows are not subject to rapid change with time. Accordingly, many surface water quality models assume steady-state hydraulic conditions. This assumption is even more valid for typical groundwater situations. For steady-state open river conditions, the product AV may be replaced by the steam discharge Q. When applied to estuarine conditions, the V and A nomenclature should be retained with V representing the net downstream velocity.

Problems in implementing Eq. (27) relate to selection of an adequate dispersion coefficient, and the postulation of tractable relations for the various a terms. For example, if the critical constituent were oxygen, one might need a term for dispersion, a term for advection, one or more terms for the oxidation of wastes, and one or more terms for reaeration mechanisms. A coefficient or reaction rate constant would be needed for all but the advection term. Determination of these constants can only be achieved by calibration, and the same criticism noted earlier for continuous flow quantity models may be raised here. In many cases the data will not allow calibration by individual mechanisms (*e.g.*, each oxygen demanding process) and thus only the lumped or aggregate effect can be calibrated.

The foregoing comments apply primarily to point source pollutant issues and to biodegradable (*i.e.*, nonconservative) pollutants. Nonpoint source loadings are usually associated with unsteady water flow conditions triggered by storm runoff. When the constituent of concern is a conservative pollutant, the primary mechanisms are a combination of dilution and dispersion. In such cases, a continuous flow water quantity model becomes the basic analytical vehicle. It must be supplemented, however, with a transport function for the determination of the nonpoint waste loadings. The data required for calibration and validation of nonpoint sovrce pollution is largely unavailable. Hence, development of a tractable and reliable nonpoint source model is a time-consuming and costly process. For these reasons the science of nonpoint source modelling remains in its infancy.

Solution of Eq. (27) in a system context varies for the constituent being modelled. In some cases direct integration is possible with reach-to-reach integrity assumed by application of continuity concepts. In other cases numerical methods, either finite difference or finite element, have been utilized. Models using Eq. (27) have been developed for and adopted by the EPA for numerous constituents including dissolved oxygen, algal biomass, biochemical oxygen demand phosphorus, ammonia, nitric and nitrate nitrogen, coliforms, radionuclides, and conservative substances.

Modelling of contaminant transport in groundwater presents some difficult questions. Laboratory-controlled experiments in a truly homogeneous and isotropic medium have demonstrated that point source pollution will disperse laterally in the shape of a 6° cone, while longitudinal dispersion will follow a normal probability distribution. Nonetheless, direct application of these results to a specific site is not without peril. A host of boundary conditions, such as, physical, chemical and biological, serve to introduce factors that may modify the overall transport picture. Variance in geologic structure as represented by stratification, localized lenticular deposits, or the existence of both lateral and vertical confining beds are examples of physical problems encountered. Process variances include ion exchange within the media, filtration and adsorption characteristics of the media, and intersection with flow from other water and waste sources. These factors are site specific. Accordingly, onsite calibration is usually required before reasonable future predictions of constituent concentration and transport is possible.

II. INSTITUTIONAL SYSTEMS OF WATER MANAGEMENT

Society imposes many demands on nature's water system. These range from providing sustenance for life to serving as a host to solitude. The breadth of these demands plus that individual members of society do not assign the same value to various water services creates many water management problems. These issues are further intensified by the migratory characteristics of the resource. Thus, mankind has found the need for governmental involvement in water resources. management.

Governmental responsibilities in water resources management can be categorized under the general headings of planning, data collection and research, legal regulation, physical regulation (development), and project operation. The relative effort in these categories necessarily varies with governmental level and legal authority. All levels engage in planning efforts which, if conducted properly, assist in carrying out the other responsibilities. The federal government can undertake major water development projects bulwarked by the commerce, general welfare, and proprietary powers set forth in the constitution, The majority of water projects, however, are operated by individuals, private sector corporations, and local units of government. A vast body of rules and regulations ranging from local ordinances to international treaties serve to govern the use, control, and protection of nature's water resources system. These rules and regulations impose constraints that water resources analysts and users must honour. What follows is a brief summary of the more pertinent features of the systems encountered in the United States.

A. Water Use Regulations

The right to use water is considered a property right in both common law and statutory law jurisdictions. Primary responsibility for administration of the use system rests with state government because constitutional interpretation places property jurisdiction at that level. Common law is the primary doctrine for most states in the contiguous United States that wholly or partly lie east of a line passing through Minnesota, Iowa, Missouri, Arkansas, and Louisiana. However, some states within this group have adopted extensive statutory codes. The states west of this line are totally dependent on statutory law or on a combination of common and statutory laws. Some states hav setatutory laws governing surface water, but retain some form of common law for groundwater. The latter mix is not conducive of reconciliation with nature's system.

Under common law, as applied to both surface and ground waters, the right to use water is based on one's location adjacent to or overlying the water source. This right, at least in legal theory, is not lost by nonuse. Early strict interpretations of the common law presumed that in the case of surface water, all common law claimants were entitled to the continued flow of their source unimpaired in quality and undiminished in quantity. These concepts necessarily had to give way to the pressures of a growing population and a growing nation. In time this early natural flow theory was replaced by a concept of reasonable use. Under this concept common law users are entitled to their domestic needs plus water for other purposes provided the latter use is of an amount deemed reasonable considering the like needs of all other users.

For the most part the common law states have also embraced the reasonable use concept of water use for groundwater. Notable, but limited, exceptions to this general rule are states that retain what is known as the absolute ownership version of common law. In the latter case a land-owner is still constrained by location (*i.e.*, must own land overlying the aquifer) but the magnitude of use is unconstrained. Upper limit constraints on use become a function of economic, geologic, and technologic considerations.

Numerical qualification of the magnitude of individual rights is never achieved under common law until and unless the issue is considered by a court of law. Findings of such adjudications are binding on only those users made a party to these legal actions.

Statutory water use law originated in the western United States. The predominant statutory doctrine is that of prior appropriation, although some of the eastern enactments depart from this. The fundamental underlying principle of the prior appropriation doctrine states that first in time is first in right. This doctrine arose in the more arid west and was adopted in an attempt to ensure water users (and investors) of a dependable continuing supply of water. Thus, an early, senior appropriator takes priority over a later, junior appropriator when there is not sufficient water to satisfy all users. This concept applies regardless of the nature of the competing uses. In jurisdictions having combined doctrines the rights developed under common law, prior to the adoption of statutory law, are designated as vested rights and are deemed superior to the earliest appropriation rights.

Statutory water law demands an extensive administrative structure. Applications for rights are received, reviewed, and granted or denied by an administratory normally titled the state engineer. The latter's actions are subject to appeal, and as an common law, the final quantitative determination of a right rests with the court. However, in the absence of any final adjudication the quantitative magnitude is presumed to be that which appears in the administrative file of the state engineer, and day-to-day administrative decisions are based thereon.

Some jurisdictions provide for market transfers of rights; others do not. However, neither initial acquisitions nor subquent transfers can be acquired and held for speculative purposes in the absence of use. Thus, an underlying principle of the prior appropriation doctrine is that rights must be placed to beneficial use. Failure to do so can make a rightholder subject to termination or forfeiture actions.

Intergovernmental considerations are complex and at times almost irreconcilable. Interstate problems have been resolved by decisions rendered in the federal court system and even by legislative enactments in Congress. Many interstate conflicts have been resolved by the interstate compact authority established by the federal constitution. These agreements are presumed to reflect the doctrine of equitable apportionment. The latter has never been defined, but assumes that the affected parties have reached an equitable apportionment or they would not be signatories to the compact. Once ratified the signatory parties must honour the allocation provisions of the compact. Similarly, international issues are resolved by treaties.

Hindsight examination of many of these documents would argue that equity is not always attained. There is also growing recognition that many single purpose compacts (*e.g.*, a water allocation compact) can hinder comprehensive water development. Accordingly, some recent compacts, such as the Delaware Basin Compact, incorporate broad planning and development powers, relate to both quantity and quality issues, and even include the Federal government as a signatory party.

The status of many water rights acquired under state law remain uncertain because of the unquantified magnitude of Indian rights associated with early treaties negotiated with the Indian Nations. The status of a number of additional rights established by state procedures also remain uncertain due to the failure to date to quantify federal reserved rights. The latter relate to judicial pronouncements that water was reserved for a variety of federal activities when public land was withdrawn or reserved for such purposes as military bases, wildlife refuges, and national parks.

B. Control of Excess Waters

When control, as contrasted to supply, of water is the problem of concern, the governing rules have roots in common law. However, the modifications are many and vary from state to state. Examination of local drianage ordinances suggests that they may also vary from city to city within a given state jurisdiction. Accordingly, legal advice is needed for most problems.

There are two relatively distinct versions of common law relating to excess waters. The Roman civil law holds that the upstream owner may exercise the advantage of his fortuitous location and is entitled to discharge water onto lower lands via natural depressions and channels. Thus, the upstream owner is viewed as dominant and the downstream owner becomes servient to his upstream neighbour. However, the upstream owner is constrained by the reference to natural channels. Thus, he cannot artifically route water down to a servient owner if that water under natural conditions would have drainage elsewhere; nor can he relocate the outlet from his land at other than the natural location.

The English version of common law embraces what is known as the common enemy rule. In this case excess water is viewed as a common enemy, and both dominant and servient owners are entitled to protect their properties from the ravages of same. Both can build dikes or levees to protect their properties from overflow.

In either version of water control law, the responsibility for damages caused must be discharged by those who altered the natural pattern of flow. Thus, a servient owner can be held responsible for backwater damages caused by his dike or levee system. Similarly a dominant owner can be held responsible for scour and overflow damage imposed on a servient owner as a result of actions that increased the natural volume of flow or changed the point of discharge.

C. Water Quality Regulation

Water users aggrieved by impairment of the quality of their water sources can seek redress under provisions of both common and statutory water use laws. Society, however, has not found these laws adequate to deal with broad public concerns related to water quality. Accordingly, a voluminous body of water quality legislation has developed during the last several decades.

This legislation has two major divisions. One body of law reletes to the development of safe potable water supplies. The second major segment relates to water pollution control. Prior to the mid-1960s primary responsibility for administering this legislation rested with the states and supplemental assistance was provided by the surgeon general of the U.S. Public Health Service. Prior to World War II the focus in both areas was on human health aspects. Following World War II the water pollution control effort gradually increased its breadth of focus. New targets included emphasis on pollution control in the interest of fish and wildlife, and the spectrium soon broadened to embrace all facets of environmental concern. The postwar, period also witnessed the modernization of a great chemical industry and the production of numerous new organics and toxics. Thus, prewar water supply, which largely focused on control of pathogenic disease, slowly but surely had to expand horizons to include appraisal of these new compounds and the health and environmental issues associated with same.

New, 40 yr after World War II, there is extensive federal legislation pertaining to both drinking water and water pollution control. This federal involvement, both regulatory and developmental, in the water quality area is fueled by the general welfare power. An initial reading of these water quality laws suggests that the federal regulatory function is very rigorous. A careful reading, however, indicates that in both fields the announced federal policy is to recognize and protect the primary responsibilities of the states to administer the controlling statutes. Current status can be characterized as retaining direct administrative powers for the regulation of water quality laws at the state level within a framework of policies and standards established largely by federal statutes and federal administrators. Individual states can, however, impose more stringent regulations than those advocated by the federal government.

In the water supply field acceptable standards for treated public supplies have been established. In the wastewater field effluent limitations have been formulated for essentially all known point source pallutants. Also, a national pollutant discharge elimination system has been installed. The latter requires the obtaining of a permit for discharging any pollutant, treated or untreated, into the public waters. Finally, in both the water supply and water pollution control fields, current law requires extensive monitoring and reporting of facility performance.

III. USER WATER SYSTEMS

User water systems are developed within the combined opportunities and constriants presented by nature's system and the prevailing legal system. The components of user systems vary according to user classifications. For example, a hydroelectric user requires no need for water treatment, whereas a municipal user would find this a subject of primary concern. User systems can be subdivided into the following components: source, treatment, distribution, return flow collection, and return flow treatment. Applicable technological concepts necessarily vary from use to use. The following discussion provides an indication of the breadth of technological issues that may arise in the implementation of user systems. Standard texts in environmental and water resources engineering should be consulted for quantitative examples.

A. Source Selection

Source selection involves a multistep procedure including

(a) estimation of projected water demand, (b) determination of raw water quality requirements, (c) identification of possible sources, (d) tabulation of legal constraints applicable to each potential source, and (e) comparative analysis of the performance characteristics of each potential source.

Estimation of projected water demands can range from a simple task to a complex problem of decision making in the face of obvious uncertainty. A small industry seeking to supply water to a new plant from a self-constructed well knows within acceptable tolerances both the average and maximum demands to be imposed. By contrast, a municipality considering the construction of a new reservoir faces appreciable uncertainty in determining the demand to be served. This is due to the lengthy time horizon associated with such projects. For example, the physical life of a properly designed reservoir will exceed 100 yr, and municipal financing laws may allow an economic life ranging from 25 to 50 yr. Attempts to forecast demand under these conditions necessarily take on the flavour of educated guestimates. Analytical approaches range from graphical extrapolation of past trends to the use of sophisticated econometric models. The economic modelling approach is widely used by higher levels of government confronted with basinwide or interbasin water development problems.

Water of acceptable quality must also be found. The determination of raw water quality requirements is relatively straightforward. In the case of public water systems the previously cited state and federal laws provide guidance. These same standards normally serve as adequate guides for industrial raw water supplies, although some industrial treated water criteria can be stringent. Raw water quality is an extremely important issue in evaluating proposed irrigation supplies. Although bacterial pollution has been of sufficient public concern that many states regulate the use of sewage effluent for irrigation purposes, the more serious irrigation quality issues relate to the presence of salts or trace metals, which can be toxic to plants. They become concentrated manyfold in the soil solution as the water is evapotranspired from the root zone. A high salt concentration in the soil solution reduces the osmotic capability of the plants and this limits their ability to absorb nutrients.

The potential salinity hazard of a source can be determined by measuring its conductivity. Generally, waters having more than 750 μmho/cm of conductivity are judged to have a high salinity hazard, although some crops will tolerate conductivities values in excess of 2000 μmho/cm. The relative amount of different salts is also important. When the sodium percentage of the exchangeable cations exceeds 10 per cent, soil permeability may be impaired and the soil will become more alkaline. Inasmuch as calcium and magnesium will replace sodium more easily then *vice versa*, the U.S. Department of Agriculture utilizes the sodium/adsorption ratio (SAR) as a means of judging the sodium hazard. It is calculated as

$$SAR = Na^+/\sqrt{(Ca^{2+}+Mg^{2+})/2} \qquad \ldots(28)$$

with the concentration of ions expressed in equivalents per million. When the conductance is 100 μmho/cm, an SAR value in excess of 18 is an indication of a high sodium risk. When the conductance measure has increased to 2250 μmho/cm, SAR values greater than 9 are considered a high risk.

Final selection among several sources, all capable of meeting quantity and quality requirements, is based on comparative appraisal of their expected performance with due regard given to applicable legal constraints. For the industry seeking to identify a specific location for its well, the decision process may reduce to a simple exercise in minimizing costs, subject to specified legal constraints concerning allowable lowering of water levels at adjacent user sites. For multisource options and for multipurpose projects, operational studies need to be conducted for representative timestreams of hydrologic conditions in order to examine reliability of performance. In cases where the same time-stream of benefits are produced, an objective function that minimizes costs is adequate. Where multipurpose benefits may be involved, the simplest decision model would maximize benefits minus costs. Care must be taken to ensure that the cost function includes treatment as well as source development costs because different sources may require different treatment processes.

For large multipurpose projects development of an acceptable quantitative objective function becomes difficult because establishing acceptable cost and benefit parameters for many environmental and social impacts is a formidable task. In these cases accepted practice seeks to present decision makers with qualitative desciptions of performance based on optimization of different objectives. For example, in a hydroelectric project one solution might seek to minimize environmental impact while another might maximize power revenues.

B. Water Treatment

Many systems require treatment before the water can be used. Governmentally mandated standards for treated water must be observed by public water systems. Industrial water use can be subdivided into the three broad categories of process water, boiler feed water, and cooling water. Process water requirements vary appreciably in industry and in some cases are far more sungent than public water system needs. Boilerfied water may also need to be demineralized. Cooling water needs usually lie between raw water and treated water standards required of public systems. Water treatment is seldom economically feasible for irrigation users. The following discussion relates primarily to water treatment as practiced in public water systems in America. [See *Wastewater Treatment.*

Major processes utilized in public water system treatment include aeration, presedimentation, regulation, flocculation, softening, filtration, adsorption, disinfection, and storage. Not all processes are needed for every system.

Aeration in water supply treatment is used to remove undesirable gases dissolved in the water. Institution, groundwater supplies, and some surface waters, contain significant amounts of soluble iron and manganese whose removal can be aided by aeration. The purpose of aeration is to provide oxygen that will oxidize these elements to a higher valance and thus reduce their solubility. Aeration devices increase the interfacial area per unit volume of water. Techniques include the use of spray nozzles, sometimes called fountains, and towers composed of a series of spills into successively lower pools or trays.

A significant segment of the treatment system is devoted to the task of separating solids from the flow stream. The processes of presedimentation, coagulation flocculation, and softening are all a part of this task. Presedimentation is not always required, but is generally applicable where the raw water contains discrete particles. The latter and particles whose size, shape, and specific gravity are relatively constant and can be removed by simple sedimentation. Other particles that can be made to aggregate or coalesce and thus change their size and shape are called flocculating particles. The design of settling basins, sometimes called clarifiers, for discrete particles is based on fluid mechanics concepts of viscous drag and the determination of terminal fall velocities.

Nondiscrete particles provide a greater challenge for the process designer. In chemistry colloidal suspensions that do not agglomerate are viewed as stable. Many raw waters have finely divided particles that do not fully meet the chemical definition, but which agglomerate too slowly for practical water treatment purposes. This difficulty is overcome by adding chemicals that assist the aggregation process. Chemicals often used to produce coagulation are aluminium sulphate. ferric chloride, or ferrous sulphate. The choice of chemical varies with the pH of the solution. Process efficiency is dependent on provision of the right dosage and by ensuring rapid mix of the chemical throughout the entire flowstream. The former is determined and adjusted day to day with the use of laboratory jar tests. The rapid mix is provided via the injection of the chemical upstream of instream impellers or a hydraulic jump. The chemical coagulants transfer the colloidal type suspensions into aggregations of particles called flocs which do become large enough to cause gravity settling. This process is further advanced by passing the chemically injected and mixed water into flocculator basins. These basins contain mechanically operated paddles that slowly stir the water to increase interparticle contact. The velocity of paddle operation is critical because too much speed will break up the floc and too little speed does not provide enough enhancement of the aggregation process. The flocculated water is then routed through one or more settling basins where removal follows the same cancepts used in eliminating discrete particles.

Excessive hardness is an undesirable element for most water use. Too hard a water leads to excessive soap consumption by domestic users. Hardness can also cause the formation of scale which will foul water heaters, hot water lines, and boilers. Certain salts produce unwanted laxative effects on humans. These problems can be overcome by softening. Softening is often required for groundwater and less frequently for surface waters. A major cost consideration in central softening relates to disposal of the large volume of sludge which is an unwanted by-product. Central softeners are incorporated into the overall solid separation efforts at the treatment plant and employ chemical precipitation. Central softening is seldom used when hardness is less than 150 mg/litre. In the latter cases the decision to soften is left to the individual user. The typical consumer softener utilizes an ion-exchange process and replaces the common constituents of calcium and magnesium with a nonhardness cation. Usually sodium serves the latter role and the interchange takes place on a synthetic resin coated with the exchange material.

Filtration is a necessary final step in the solids separation. It serves to extract the small particles that are not removed by settling. Filtration also eliminates many pathogenic organisms. Filter design as practiced in the latter half of the twentieth century utilizes granular media beds ranging from 2 to 2.5 ft in depth and flow rates of 2 to 8 gal/min/sq ft surface area. Filters become plugged with the material they extract from the flowstream. If the flow of liquid is reversed through the filter; these deposits can be loosened and flushed away to wastewater drains. The process of reversing the flow through the filter is called backwashing and serves to rehabilitate the filter media. Filter efficiency can be enhanced via the use of mixed or dual media having different sizes and specific gravities. Properly selected mixed media will produce a situation where the larger but lighter media is redeposited at the top of the media bed following the backwash cycle.

Development of water treatment systems was initially motivated by the need to minimize waterborne disease. Accordingly, the use of disinfectants has been a long established step in treatment systems. In recent years chemical, as contrasted to bacteriological, contamination has become of increasing concern. Many of the new chemical compounds produced by humans have been shown to be carcinogenic and/or otherwise toxic. The long-term health effects of many of these compounds is not known.

The initial use of disinfectants was spawned by the desire to eliminate pathogenic microorganisms. Studies indicate that 90 per cent or more of the bacteria and viruses in the water are removed by the combined solids separation and filtration processes. Disinfection has been the primary tool for finishing this task. Thus, the disinfectant is normally added after the water leaves the filter. Chlorine has been the primary chemical used in the United States. Over the years disinfectant has also been added to the treatment process for other purposes. In some plants it is applied as prechlorination to the incoming flow to assist with the oxidation of inorganics (presedimentation) or to eliminate biological action that produces undesirable gases. Chlorine has also been added prior to filtration to prevent the growth of bacterial colonies on or in the filter. However, the major use of chlorine has been for pathogenic protection in the treated water and as a source of continued protection against bacterial growth in the distribution system.

Recent studies have shown that chlorine will oxidize some of the new organics and produce undesirable products; basically chlorinated hydrocarbons and haloform compounds, some of which are suspected to be corcinogenic. Accordingly, the search goes on for new and more stable disinfectants. Non-chemical processes include irradiation with ultraviolet light, electrocution, heating, and even sonification. Ozone is often used in European systems. Ozonation is significantly more costly than chlorination and it does not provide residual protection against bacterial growth in the distribution system. However, ozone does not present the problem with the oxidized organics, and thus its use may increase.

Alternatively, modifications in current chlorine methodologies may prove adequate. Research indicates that substitution of chloramines for free chlorine will minimize a number of the problems encountered to date. The chloramines do not react with organics as does free chlorine. Still another alternative utilizes adsorption processes to remove the offending organics. The adsorbent used to date had been powdered carbon. Costs are high and research on adsorption and other processes is sure to continue.

C. Distribution Systems

Distribution system requirements like the other components of the user system vary appreciably. For a small industrial user it may be limited to the installation of a one building piping system. For the public water system it will normally require a complex set of pumps, transmission mains, system storage, and distribution pipes plus appurtenant valves, hydrants, and metres.

The typical water distribution system utilizes pumps to force the water through the pipe system from the treatment facility to the direct user. This pumped flow is usually supplemented by one or more distribution reservoirs located near high demand centers and in such a way as to be able to contribute added flow during emergency periods via gravity release. These distribution reservoirs also ensure the maintenance of adequate pressure in the system. Distribution storage requirements for the reservoirs are determined by mass balance analysis throughout the appropriate time cycle (usually one day) and for selected operational conditions which normally incorporate maximum fire requirements.

Pipe distribution systems should be constructed in grid form as constrasted to a branching or tree type system. Use of a grid avoids dead end lines that can spawn bacterial growth in the stagnant flow areas where an adequate disinfectant residual is difficult to maintain. Moreover, with a grid system emergency repairs can be made with minimum impact on other connections. Finally, a grid system allows for maintenance of an adequate pressure regime around the loop.

In U.S. practice the adequacy of the water distribution system impacts directly on fire insurance rates. Accordingly, governmental regulations, usually local, specify by code the size of pipes permitted for different distribution functions, the minimum pressures to be maintained, the size and specing of hydrants, and other physical dimensions of the system.

In earlier days analysis of the hydraulic capability of a distribution system was a labour-demanding task. This is no longer true. Most existing systems now utilize or have consultants who utilize digital programmes designed to examine system performance for a wide range of operating conditions. These programmes provide simultaneous solution of a set of equations developed for each segment of the system. For each pipe segment the input data requires a listing of the demand to be served and a statement relating the required energy head (E_H) to the flow rate (Q). This relationship is of the following form

$$E_H = CQ^m \qquad \ldots(29)$$

and is a simplification of the basic Darcy-Weisbach equation for flow in a pressure conduit. Thus, the values of C and m for each segment reflect the specific friction, length, and diameter characteristics of that segment. Given this input, the programmes solve for flow and pressure distribution throughout the system subject to the following constraints. First, the flow into each junction must balance the flow out of the junction. Second, the summation of E_H around any loop in the grid must equal zero.

D. Used Water Collection Systems

Some water uses, such as stream recreation, do not divert water from the source. Such instream uses are nonconsumptive. Other uses that do divert water from the stream are viewed as consumptive. The amount of consumption, however, seldom equates to the amount of diversion. In irrigation use the difference between diversion and consumption is called return flow. In public and industrial water systems this difference is termed wastewater. In both instances these flows represent used waters. These flows are subject to legal regulations. In highly developed irrigation areas in the western United States one can obtain rights to the use of return flows. Throughout the country the quality of industrial and municipal wastewaters, and to a lesser extent irrigation return flows, are being subjected to increasingly stringent quality constraints. User systems, therefore, face responsibilities for the collection and treatment of these used waters.

The initial question posed is how can the flows be collected. In irrigation this question is answered by the installation of a field drainage system unless there is opportunity for adequate percolation to a self-draining saturated groundwater strata. The need for drainage in irrigation presumes the application of water in excess of plant needs. Excess water is supplied to avoid a salt buildup in the root zone. These excess waters are used to leach the salts out of the root zone and the drain is used to convey the leachate offsite. The need for drainage also occurs if a subsurface confining layer leads to the development of a perched and rising groundwater table spawned by the leachate. Too high a water level will swamp the roots and again impair agricultural production. The return flow necessarily contain a higher concentration of salts than was contained in the irrigation waters. Moreover, in relatively arid zones the leachates pass through soils not fully flushed by natural precipitation and thus have ample opportunity to increase their mineralization. The quality of agricultural return flows, long ignored, is becoming of increasing concern in modern water management.

For too many years some industrial and municipal waste waters were collected via a piping system and discharged untreated into receiving streams. Imposition of water quality regulations has essentially forced the elimination of untreated wastewaters. Wastewaters are collected via a complex system of pipes somewhat like a water distribution system, but which have significantly different physical properties. The pipes that carry wastewaters are known as sanitary sewers.

The foremost difference between a wastewater collection system and a water distribution system is that the former uses gravity flow wherever possible while the latter uses pressure flow throughout. Wherever gravity discharge is not possible—for example, in bringing the wastewater from one small watershed across the topographic divide to another basin that offers a better route to the treatment plant—a forced (pressure) main is used. The motivation for use of gravity flow is twofold. First, it allows one to eliminate the costly energy charges of a pressured system. Second, gravity flow conduits are not designed to run full and thus, in conjuction with manholes and other installed vents, provide a ventilation system, which is desirable for two reasons. First, wastes contained in the flow can develop very explosive mixtures of gases and or very odourous conditions: The venting minimizes the opportunity for serious damage and the development of an excessive nuisance condition. Second, in the absence of an adequately ventilated system, maintenance employees face a real threat of becoming asphyxiated.

E. Wastewater Treatment and Disposal

The Water Pollution Control Act of 1972 provided the basis for establishment of wastewater discharge standards. Prior to the adoption of that act requirements in many states were based on the relative appraisal of the waste load and the transport capability of the receiving stream. Now all wastewater dischargers are issued a permit that specifies the amount of each pollutant that they can discharge. The entire range of potential contaminants are subject to regulation via this permit system. Contaminants now targets for treatment include pathogens, radioactive wastes, dissolved inorganic solids, biodegradable organics, refractory organics, nutrients, suspended solids, metals, and heat. Almost all unit operations and processes utilized by the sanitary and chemical engineering industries can find application in wastewater treatment if one seeks total elimination of all wastes.

Wastewater treatment is classified as primary, secondary, or tertiary with primary being least effective and tertiary most effective. Primary treatment is essentially an exercise in the removal of a portion of the suspended solids, primarily by sedimentation, and the subsequent disposal of the deposited sludge.

Secondary treatment supplements the sedimentation efforts of primary treatment with biological processes to eliminate the dissolved and colloidal organics in the wastewater stream. This is accomplished by placing colonies of microorganisms in contact with the organic waste. These organisms convert the organics into biomass which can then be settled and removed as sludge. Secondary treatment usually meets the requirements of discharge permits, but for some industrial wastes compliance cannot be achieved with secondary treatment. Secondary treatment is generally expected to reduce effluent biochemical oxygen demand to 30 mg/litre or less. Suspended solids are also expected to be lowered to the 30 mg/litre or less level. Secondary treatment does not provide an equal level of removal of nutrients.

Tertiary treatment is called upon when further removal of suspended solids is necessary and/or if nutrients appear to pose an unusually serious threat. The additional solids removal can be achieved via filtration techniques regularly employed in water treatment plants. Nutrients and other hard to treat wastes can be removed by a variety of biological, physical, and chemical processes. The latter can range from land treatment systems (*i.e.*, the disposal of wastes on the land with subsequent uptake of the nutrients by plants) to costly carbon adsorption units.

A critical aspect of wastewater treatment relates to the problems of handling, disposing, or otherwise managing the sludge produced at all levels of treatment. Paradoxically, disposal of sludge, which is the by-product of wastewater treatment, is a major waste control problem. As much as 50 per cent of both the construction and operation costs of wastewater treatment plants can be devoted to sludge handling and disposal.

A final decision for any user system is the question of where to dispose of the treated wastewater. Increasing attention is being directed to the need for water reuse. In much of this country and in other countries around the globe, indirect reuse, even for potable water supply, has been practiced for many years. That is, treated water has been returned to a stream and then been withdrawn by another user downstream. Direct reuse in the sense of recycling the treated effluent to the water supply intake of a potable system has not been practiced except as an emergency measure. Although the trend is toward increasing reuse, the recent appearance of chemical substances that are difficult to measure, even more difficult to remove, and of unknown health risk has served to slow the advocacy for direct reuse.

There are many alternative ways of disposing of treated wastewater. The most common is the direct return to surface streams where the wastewater is further diluted and available for various downstream uses. Land application treatment processes automatically include the disposal decision. Treated wastewater can also be deliberately used for irrigation and/or some industrial uses such as cooling. Treated wastewater has also been used to recharge or otherwise stabilize groundwater levels threatened by overpumping and/or saltwater intrusions. The proper disposal choice is site specific and not necessarily based solely on economic considerations. To illustrate, use of land disposal treatment schemes may run afoul of established downstream water rights. The seasonal variance in irrigation water need may differ from the seasonal pattern of wastewater generation. Moreover, for both irrigation and industrial use the nature of the remaining waste products in the flow can be most critical. An all inclusive generalized solution cannot be given. The user must remember that alternative choices may exist, and all should be evaluated before a final decision is made.

BIBLIOGRAPHY

Bras, R.L., and Rodriguez Hurbe, I. (1978). "Random, Functions and Hydrology." Addison-Wesley, Reading, mass.

Cunge, J.A., Holly, F.M., Jr., and Verway, A. (1980). "Practical Aspects of computational River Hydraulics." Pittman, London.

Goodman, A.S. (1984), "Principles of water Resources Planning," Prentice-Hall, Englewood Cliffs, N.J.

Linsley, R.K., and Franzini, J. (1979). "Water Resources Engineering." McGraw-Hill, New York.

Linsley, R.K., Kohler, M.A., and Paulhus, J.L.H. (1982). "Hydrology for Engineers." McGraw-Hill, New York.

Loucks, D.P., Stedinger, J.R., and Haith D.A. (1981). "Water Resources Systems Planning and Analysis," Prentice-Hall, Englewood Cliffs, N.J.

Major, D.C., and Lenton, R.L. (1979). "Water Resources Systems Planning." Prentice-Hall, Englewood Cliffs, N.J.

Peavy, H.S., Rowe, D.R., and Tchobanolous, G. (1985). "Environmental Engineering," McGraw-Hill, New York.

Steele, E.W., and McGhee, T. (1979). "Water Supply and Sewage." McGraw-Hill, New York.

Yevjevich, V. (1978). "Probability and Statistics." Water Resources Publications, Ft. Collins, Colo.

4

THE EVALUATION OF BIOLOGICAL RESPONSE BY TOXICITY AND WATER QUALITY ASSESSMENTS

A. The Environment

1. Optimum conditions

All living organisms exhibit a preference for the environmental conditions in which their life processes take place most efficiently. In neritic, or coastal waters these environmental conditions are, in general, so variable that relatively few of the inhabitants are ever likely to experience all these preferred conditions simultaneously and most function perfectly well when marked departures from the desirable, *i.e.*, the optimum, occur. In contrast, the oceanic environment, and especially that at great depth, is more nearly uniform and the inhabitants, of these areas must be considered as living in near optimal conditions: thus they are more susceptible to marked alternations in water quality than the neritic species.

The differing responses of oceanic and neritic organisms may be domonstrated by reference to their relative ability to osmoregulate, *i.e.*, to control the concentration of their body fluids. Animals may be homoiosmotic or poikilosmotic: the former maintain their body fluids at concentrations differing from that of the environment, whereas the latter are in osmotic equilibrium with the surrounding medium. In the open sea most invertebrate species are poikilosmotic, though fish in general are not; on the other hand, these organisms which live in the more variable estuarine environment tend to belong to the homoiosmotic group (Allee *et al.*, 1949). However, species such as the lugworm Arenicola marina, that are poikilosmotic, do nevertheless penetrate those regions of an estuary where significant reductions of salinity occur during some part of the tidal cycle (Beadle 1972): in such cases the mode of life assumed by the particular animal, *e.g.*, the burrowing habit of Arenicola, coupled with an ability to tolerate anaerobic conditions (Patel and Spencer, 1963) and thus refrain from irrigation of the burrow when conditions are adverse (Wells, 1949), enables these species to live in otherwise unsuitable conditions. Nevertheless, those poikilosmotic species and phyla which lack both an effective mechanism for the control of osmotic and ionic change and a behavioural mechanism whereby immunity from the influence of a much reduced and fluctuating salinity is confered, *e.g.*, the echinoderms, which include the starfish and sea-urchins, connot colonize these waters and are absent from fresh waters (Allee *et al.*, 1949).

These different responses to variations in salinity are consistent with the general statement that the ability of animals to tolerate a departure from optimal environmental conditions, including temperature, depends upon a complex relationship of the interacting variables. These include:

(a) the species or other taxonomic division,

(b) the external environment in which a given population normally lives,

(c) the duration of exposure to adverse conditions,

(d) the "milieu interieur" which is more variable than the original concept suggested,

(e) the rate of change of internal conditions as extreme conditions are approached, including the influence of behavioural and physiological adaptations and

(f) the general condition of the animal with regard for example to thermal insulation, state of health, infection by disease or parasites, and the stage in the life history (Allee *et al.*, 1949).

2. *Single variables*

Consider, now the response of one species to a single environmental variable. One might reasonably predict that processes that are at their most efficient and most active at the optimum level undergo a regular decline in efficiency when subjected to progressive excess or dificit; in other words, the response evoked would be described by a curve symmetrical about the optimum. While this may be approximately true for a number of species, some asymmetry is more usual and occasionally the response curve may be strongly skewed. The fact that not all the species of a particular genus, much less all the members of the same phylum, evince precisely similar response curves may be seen with reference to the dinoflagellates (Braarud, 1951; Nordli 1957; Fig 4.1; moreover, the time of collection from the wild may influence the laboratory response of a particular species. Furthermore, if the experimental conditions are changed by the use of different batches of sea water, different light conditions or different temperatures then the form of the response curve may change also. Where a response curve is strongly skewed, it follows that the optimum condition is close to coma and death as one or other environmental extreme is approached. In particular, both warm-and cold-blooded animals operate most efficiently at temperatures near to the maximum of toleration, often optimum temperature and the point at which heat coma and death ensue are close. The safety factor is greater at the lower than at the upper end of the tolerated temperature rang (Allee *et al.*, 1949). Evidently, animals that live in a temperate climate are in a more favourable position with respect to the disposal of waste heat than are the tropical species. For the latter, ambient, optimal and lethal temperatures may be little different (Reeve and Casper, 1970).

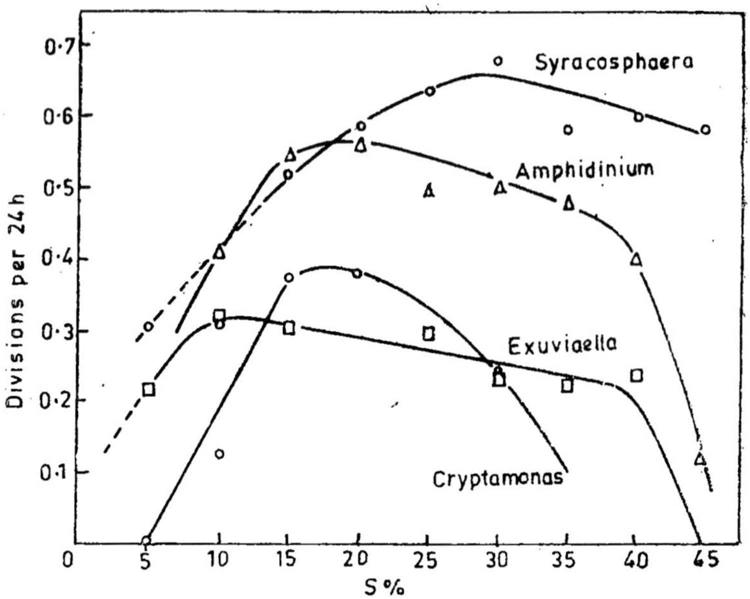

Fig. 4.1. Salinity response curves for Syracospbaera carterae, Cryptomonas sp., Amphidinium sp. and Exuviaella baltica after Braarud, 1951).

3. *Multiple variables*

The overwhelming majority of experimental studies on the response of marine animals to environmental variables has been concerned with single variables, and not as customarily occurs in the natural environment with multiple factors (Alderdice, 1972). Thus the concentration of O_2 in sea water is inversely related to temperature and salinity (Truesdale and Knowles, 1965; Gilbert *et al.*, 1967) and at any given temperature survival of fish is related to O_2 concentration (Davison *et al.*, 1959). It is generally considered that the toxicity of any material dissolved in water is approximately doubled by a rise in temperature of 10°C: an effect which Swedmark *et al.*, (1971), for example, demonstrated with respect to the effect of nonyl phenyl ethoxylate on the cod and flounder. At a temperature of 6-8°C, they found a 69h LC_{50} of 6 ppm for both species, but at 15°C it was 2.5 and 3 ppm respectively. Multifactorial studies are interpreted by means of response surfaces instead of a family of curves as shown above. Such studies, although frequently concerned with the influence of salinity and temperature in combination, have revealed that not all activities of the particular animal are optimal for any one salinity-temperature combination.

Within limits, the multifactorial technique has rather more to offer than that which yielded, but as already stated, its use has been limited both in frequency and scope primarily to salinity-temperature combinations. However, the sea presents many times two factors working in combination and it seems that a truly multifactorial analysis has great potential value, particularly in relation to natural optimum conditions and the influence of one or several discrete pollutants. There are formidable difficulties that beset this approach, however, for accepting that while the means for multifactorial analysis may be available, the sheer scale and technical difficulty of aquarium work would probably render them too costly for general use, more particularly with respect to studies lasting a year or more. Moreover, use of such a refined technique on such a scale would imply that the difficult art of long-term maintenance of marine organisms matches such refinement. Problems such as the unseen influence whereby community tanks, apparently constructed, stoked and handled exactly alike, produce, if left alone, very different biota, are not understood and seem impossible to overcome. Therefore, while noting that the use of such methods must eventually increase, they are likely to make little contribution to studies of biological response in the near future and regretfully the are excluded from further consideration below.

4. Essential substances

Thus far, the general relationships between the living organism and natural parameters of its environment have been considered. This approach is pertinent since by far the major components of all wastes are nutritive materials required by biological systems. Of these, sewage is the most outstanding, both in total amount and in its effect, for it contains materials such as nitrate and phosphate, which are essential to the well-being of the biota as a whole. However, in excess, sewage, because of its high content of organic matter, depletes the ambient O_2 supply with inevitable adverse consequences. In the same way, trace metals such as Fe, Zn, Cu and Mo, for example, are vital to the health and well-being of the biota, but are toxic in excess. Clearly, then, it is vital that such materials be supplied to the biota in proper amount and must be recycled if a healthy system is to be maintained. Consequently, where these essential materials are involved, the photosynthesizing biota will have optimum requirements for each, and deficiency or excess of any acting either independently or in combination will have adverse effects on the efficiency and productivity of the biota. Thus, two of the fundamental principles of water quality management are:

(a) recognition that biological systems work in this way, and identification of the key plants and animals, and

(b) understanding specific plant and animal communities whereby such materials may be removed and recycled through the environment without creating such a departure from optimum conditions in the receiving water body that the zone of tolerance conditions of this community are disrupted and the body of water becomes unbalanced, polluted or poisoned (Perkins, 1974; Fig. 3.2).

B. The Organism

1. Organism Interactions Including Non-predatory Relationships

The foregoing paragraphs have been concerned with the effects of certain parameters in the chemical and physical environment on the biota; however, important though these influences are, it should not be forgotten that living organisms also interact with each other. The most obvious of these interactions is that

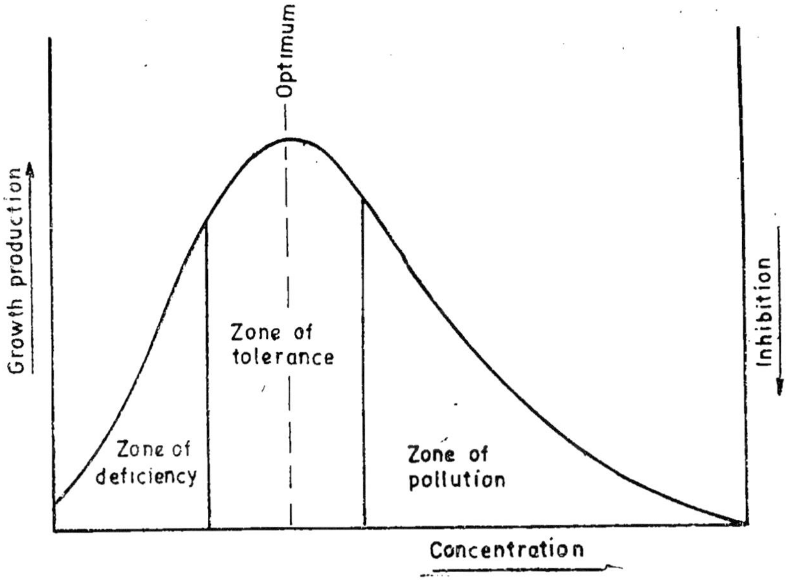

Fig. 3.2: The relationship between the concentration of growth factors, optimal conditions and the development of deficient and polluted environments (after Perkins, 1974)

herbivores graze on plant life and that carnivores prey upon herbivores. There are, however, more subtle interactions within the biota that have similarly far-reaching consequences. For example, most rocky shores are colonized by algae which support large numbers of the winkle Littorina, together with other herbivorous species and carnivores which prey upon them. Normally the algae and the fauna dependent on them are in approximate equilibrium and the algae flourish well enough to provide a more or less thick cover of the rock surfaces. This approximate equilibrium, however, may be relatively permanently disrupted. For example, in the autumn of 1962 a householder at Garelochhead stripped an adjacent shore of its brown algae (nearby colonization would suggest that this included Fucus spiralis, Ascophyllum nodosum and Fucus vesiculosus) to use as garden manure. Since then there has been little effective colonization of that shore by algae, because the winkles remained in such abundance that any new settlement of algae has been rapidly eliminated. Clearly, a dynamic equilibrium subject to such violent disruption may require a prolonged period for its restoration: here, after 14 years, there are no signs that it is imminent or likely. Similarly, the death or displacement of the herbivore populations on shores affected by the Torrey Canyon oil-spill permitted an exceptional growth of the alga Enteromorpha (Smith, 1968; Nelson-Smith, 1968) for a short time.

These are extreme examples which resulted from human intervention. Nevertheless, even in situations where all the component species flourish and all are apparently living in conditions that permit normal growth, high productivity and effective reproduction, it is not uncommon for the ecologist to observe marked discontinuities and a general non-homogeneity of distribution in the component species. In part, such discontinuities arise from the choice of habitat and the behaviour evinced by larval and adult animals (Meadows and Campbell, 1972). The preferences exhibited may be for a physical, biological or chemical attribute of the surface chosen, and more than one of these factors may operate simultaneously. For example, the cyprid larvae of the acorn barnacle exhibits a rugophilic response, *i.e.*, it tends to settle along and within grooves that are either much smaller or considerably larger than itself (Crisp and Barnes, 1954). However, these larvae also exhibit a marked gregariousness in their settlement (Knight-Jones and Stevenson, 1950; Crisp and Knight-Jones, 1953). Where smooth glass panels are used as settlement plates, Elminius settles more readily if the individual plate is already colonized by recently settled barnacles, or if a barnacled microscope slide is attached to the glass plate. Moreover, settlement is always greater on those plates that are exposed near rocks colonized by barnacles than on similar plates exposed near areas of bare mud only 50-100 m away (Knight-Jones and Stevenson, 1950). When settling near other barnacles, cyprids apparently make no specific recognition of spat of their own species (Knight-Jones, 1955; Meadows, 1969), but barnacles of other families have no appreciable stimulatory effect (Knight-Jones, 1955). Cyprids may respond positively to bases of dead barnacles, but an abundance of older barnacles may make some cyprids leave a surface without metamorphosing (Meadows, 1969). Moreover, the cyprid larvae in making a gregarious settlement require to make physical contact with settled barnacles or their bases, which contain among other components an insoluble quinone-tanned protein formed from anthropodin chemically cross-linked with polyphenols, to which they react by a tactile chemical sense (Knight-Jones,1953; Crisp and Meadows, 1962, 1963). Clearly, such physical, chemical and biological processes have little to do with the essential metabolic requirements of the juvenile or adult animal or with pelagic food provided by the environment at large, but they are vital determinants in the colonization of a given situation. The present or recent past existence of its kindred species is a prerequisite for the settlement of such species.

The important role which the nature of the substratum plays in relation to the distribution of polychaete worms has long been recognized (Day and Wilson, 1934; Wilson, 1937, 1953). The pelagic larva of Opbelia is able, under favourable conditions, to settle and metamorphose immediately upon contact with a substratum suitable for adult life. Failure to make contact with a suitable sediment delays metamorphosis until one more suitable is found; in its absence, the larva eventually dies either without metamorphosing or attempting to do so. In a classic study, D.P. Wilson was able to show that attractive and repellent factors existed on sand, that the attractive factor was associated either with micro-organisms living on the sand grains, or with the organic products of these organisms, and that only such sediment particles were likely to attract these larvae (Wilson, 1953a, b, 1954).

Near and for some depth below the low-water mark in the exposed outer areas of temperate estuaries, massive colonies of tubiculous polychaetes of the genus Sabellaria create large hummocks with rocks as a base, or they spread over the sand from some relatively secure base, thus consolidating large areas of otherwise mobile sand. In the Solway Firth such colonies, known as "the Ross", extend for many miles along the south shore in a zone also colonized by Lanice conchilega and elsewhere Owenia fusiformis (Perkins, 1973). The capacity of a Sabellaria colony for gathering food and building materials is proportional to the surface area of the colony (Wilson, 1968). It is, therefore, a distinct advantage if its pelagic larvae settle in close relationship with established colonies and with each other. As with barnacle cyprids above, and other polychaete larvae, the Sabellaria larvae have a searching phase prior to metamorphosis upon a suitable substratum. This metamorphosis is not greatly influenced by physical factors and the strongest stimulus to settlement is received by the larvae when they make contact with the tubes of adults of their own species, with their remains, or with the tubes of recently settled young. The essential factor in this settlement is the cement secreted by adults and young alike with which the tubes are constructed; direct contact with this substance is essential since it cannot be detected from a distance. This metamorphosis-inducing substance in the cement of sandy tubes and in the secreted primary mucoid tubes is insoluble in water and is unaffected by drying. Larvae which have been aged in clean, glass vessels become progressively less capable of secreting a fully formed mucoid tube when placed with adult-tube sand; eventually they become unable to secrete even a partial tube. Later, they may undergo an imperfect metamorphosis or die without metamorphosing. However, some larvae successfully undertake metamorphosis, upon which the other larvae that make contact with them quickly do likewise. The ability of some larvae to metamorphose when adult colonies are absent is clearly an advantage to the creation of new colonies (Wilson, 1968). The larvae of both Sabellaria spinulosa and S. alveolata are able to survive initial periods of starvation and, provided the periods are not too long, can complete development apparently unaffected. Both detect by direct contact the tubes of their own adults, which can act as releasers inducing settlement. The settling larvae can distinguish the tubes of the adults of their own species, but S. alveolata may, under some circumstances, choose the tubes of S. spinulosa instead of those of its own species (Wilson, 1970a, b); the larvae of S. alveolata are little influenced by the mineralogical character of the tube walls (Wilson, 1970a).

The serpulid polychaetes secrete calcareous tubes by means of which they attach themselves to a variety of substrata depending upon the individual species, many of which have a world-wide distribution (Nelson-Smith and Gee, 1966; Crisp *et al.*, 1967). The spirorbid Spirorbis borealis is characteristically an inhabitant of fucoid algae, upon which it may occur in large numbers. Its larvae are liberated mainly about the time of the moon's quadrature. Upon liberation, these larvae are positively phototactic for up to 2 h, after which they swim at random for a further period of up to 2 h, during which they visit a large number of different surfaces, crawling upon them and swimming off again. Once a suitable surface is chosen, they secrete an initial semitransparent tube and metamorphose. These larvae will settle on a glass or stone surface only if it is colonized already by a bacterial film, but they have a strong preference for the surface of the brown seaweed Fucus. They are so markedly gregarious, that concentrations of recently settled individuals can reach 10-20 cm^{-2} (Knight-Jones, 1951). When a number of the larvae of S. borealis are freshly liberated the larvae show a marked gregariousness. However, after they have been kept swimming for 3 h less choice is exercised and larvae settle rapidly both in isolation and group-wise. After being kept swimming for 8 h, the rate of settling declines and many larvae are apparently too weak to settle successfully (Knight-Jones, 1953b). On the coast of Wales, at least, Spirorbis borealis occurs most abundantly on Fucus serratus. Studies of the settlement of its larvae on "Tufnol" panels, treated with an extract of F. serratus, revealed that the larvae selected extract-treated surfaces in significantly greater numbers than the controls, indicating that settlement is a response to some chemical substance derived from the alga, rather than to changes in the physical properties of the surface. This reaction results from the ability of the larvae to detect the active substance as an adsorbed layer on a solid substrate rather than as a concentration gradient in the water. The precise mechanism of the response is unknown but, compared to larvae in contact with untreated surfaces, those in contact with the substance undergo a curtailment of the photopositive and crawling behaviour evinced prior to settlement (Ryland, 1959; Williams, 1964). Another spirobid, S. rupestris, normally lives on rocks encrusted with Lithophyllum polymorphum. The larvae of this polychaete settle very readily on L. polymorphum, but not on other coralline algae or on fucoids (Gee and Knight-Jones, 1962; Gee, 1965). They will settle rapidly on slates only if they have been first soaked in an aqueous extract of L. polymorphum; the active agent will pass a dialysing membrane of average pore diameter 24°A, but its effectiveness is reduced by treatment with heat and formalin (Gee, 1965).

Where a particular habitat is potentially suitable for the settlement of several species then, in addition to the act of settlement itself, other more complex interactions such as "competition" may become apparent. J.R. Lewis has been concerned with a long-term study, on the north-east coast of England, of the rocky shore populations of acorn barnacle, Balanus balanoides; mussel, Mytilus edulis; limpets, Patella vulgata, P. aspera; dog-whelk, Thais lapillus; and brown seaweed, Fucus spp. He found that at some sites the populations have remained stable over long periods and are composed of slow-growing, long-lived individuals maintained by low recruitment, others situated in apparently similar habitats are subject to dramatic fluctuations on a seasonal, annual or irregular time scale. In general, stability and longevity are inversely related to settlement, growth and predation rates and increase progressively upshore. Certain chains of events became apparent.

1. The 1965 class of Thais ceased growing during winter and were eaten in great numbers by the Purple Sandpiper Calidrys maritima. The following winter was mild, the 1966 year class fed and grew continuously and became too big for Calidrys to eat. The increased number of larger, juvenile Thais (compared with 1965/66) ravaged the mussels and made space available for barnacle colonization. Areas, which, by chance, lacked Thais retained their mussels and so by July both barnacle and mussel-dominated areas existed side by side, a

condition stemming primarily from the influence of temperature upon juvenile Thais, six months previously.

2. Extremely heavy Mytilus settlement (1967) destroyed all other species locally, but severe competition for space caused unstable hummock formation and eventual destruction by severe storms in March 1969. Space was now available for the barnacle settlement due in June. However, the mussels had excluded limpets since 1967 and had not themselves been destroyed until after the current limpet settlement period. Accordingly, diatoms and ephemeral algae developed abundantly and quickly, and barnacles failed to establish.
3. Experimental removal of limpets among barnacles led to an immediate growth of algae. This caused increased mortality of existing barnacles and greatly reduced barnacle settlement. Limpets migrating on to bare rock beneath the algae grew more rapidly than those among barnacles, preventing recolonization by further algae and so promoted reoccupation by barnacles. (Lewis, 1970.)

More subtle still are the non-predatory relationships mediated by external metabolites, a concept developed by Lucas (1938, 1947, 1955, 1961a, b) and recently reviewed by Johnston (1972). In essence this concept postulates that in its lifetime an organism, plant or animal secretes organic materials that may be of advantage or disadvantage to associated organisms. Where these substances are of advantage, they may be growth factors or may act as a trigger which ensures that a particular succession of organisms can take place. Since that first expression of the significance of external metabolites (which may be hormones, antibiotics or vitamins), the importance of these substances to life in the sea has become realized generally. For example, some species of Protista have a requirement for vitamins and the yield of one, Monochrysis lutheri, is directly proportional to the concentration of vitamin B_{12} at concentrations between 0.1 and 100 $\mu\mu$g/ml (Droop, 1958, 1961). Compared to the open sea, the waters of fertile bays are rich in amino acids (Park *et al.*, 1963). The salt marsh is an important source of supply (Starr, 1956; Jones and Stewart, 1969a, 1972), as indeed is the runoff from the land (Stewart, 1972); vitamin B_{12} is produced by prokaryotic bacteria, blue-green algae (Stewart, 1972) and fungi and amino acids by blue-green algae (Jones and Stewart, 1969b). On the other hand, increased amounts of carbohydrate and organic matter in the supporting water accompany an increase in diatom numbers (Marshall and Orr, 1962, 1964; Marker, 1965). Khailov and Erokhin (1971) showed that the copepods Tigriopus and Calanus can accumulate both monosaccharides and polysaccharides from sea water and that external metabolites may comprise up to 13 per cent and 1.2 per cent respectively of the total food intake of these animals. So far, the external metabolites discussed are clearly beneficial, or at least neutral, in effect. However, neurotoxic secretions produced by the planktonic "red tide" organisms, *e.g.*, the dinoflagellates Gonyaulax polyedra and Gymnodinium breve, can be fatal to Man and animals: shellfish become affected and enormous numbers of fish may die in the presence of such blooms (Nightingale, 1936; Davis, 1948; Needler, 1949; Ray and Wilson, 1957).

To summarize, the relationship between the living organism and its animate and inanimate environment is complex, and the interdependence of the individual members of the biota is much more important than is disclosed by the consideration only of food and prey-predator relationships. Clearly, the interactions revealed by the work of Lewis (1970) and the influence of non-predatory relationship mediated by external metabolites suggest that every component species of an assemblage of the biota must function appropriately if the whole is to be healthy and productive. Given this conclusion, it is evident that the view, so widely canvassed, that the loss of one species of no economic value has no significance to the wider communities of the sea is fallacious, since for good or for ill it has an influence on the health and productivity of the whole community, including the species of economic importance. Consider an analogy: a knitted garment, when intact, is both durable and comfortable, but soon becomes unravelled to the wearer's discomfort if one broken loop remains unrepaired. This analogy is too simple, because the loss of one species does not inevitably kill or completely disorganize the whole community since often the impact of the loss is absorbed by the system. Sometimes, however, the missing organism may play a key role for a transitory stage of fish or shellfish of economic importance. Such species may not be satisfactorily replaced by the organism moving into its niche. Nevertheless, the implications are clear, improvidence or bad management possibly leads to irretrievable loss, and the information currently available is insufficient to determine what free-living species may or may not be safely dispensed with.

Further difficulties arise because the sea is a fluid medium which bathes all its inhabitants and through which, eventually, all soluble and insoluble materials become dispersed. Unlike the land, where materials may remain at the site of introduction for prolonged periods, substances introduced into the sea may ultimately influence a wide area and by some asymmetry of biochemistry or sediment transport may become highly reconcentrated at some site far distant from the point of introduction, Clearly, if the second condition, for the maintenance of a water quality within a zone of tolerance about the optimum is to be satisfied, the influence of introduced wastes on the biota as a whole must be under-stood, and the influence of introduced toxicants must be understood to the point that no gross area of uncertainty remains. This statement entails that not only must the effect of acute exposure be understood, but that studies must also include truly sublethal clinical studies, and not just investigations over a range of apparently sub-lethal concentrations (meaning sub-acute only) to which, on prolonged chronic exposure to the toxic entity, significant or total mortality may result. Only by these means may effects in relation to either productivity and the food chain or survival to successful reproduction be assessed, for it is only in these vital terms that studies in marine toxicology can have any practical meaning in relation to the wider communities of the sea, including marine resources.

I. BIOLOGICAL RESPONSE AS MEASURED BY TOXICITY TESTS

A. The Problem

The essential problem which faces the would-be marine toxicologist is the identification of those members of the population that are (a) most susceptible to an introduced toxin and (b) likely to have most influence if they become either debilitated or removed from the community. This matter of identification of the core of the problem and the means by which it may be adequately examined is nearly as complex as the biological relationships in the sea itself for the reasons noted above. It follows, therefore, that to measure toxicity productively one must be quite sure that the organisms or organisms used are not only convenient in the sense of availability in large numbers and relevant to the particular situation under review, but that a suitable cross-section of sensitivity to intoxication is chosen. In other words, that the influence of selective toxicity has been adequately recognized. Furthermore, working on a particular organism without the aim of management of a resource has little contribution to make towards the solving of real problems.

This emphasis on selective toxicity stems from the fact that all toxic materials are not equitoxic to all organisms, selective toxicity is much used in medicine and agriculture (*e.g.*, Albert, 1968) and applies equally to marine organisms. Moreover, the ranking of species sensitivity to toxic entities may alter depending upon the type of toxin involved, for example acorn barnacles are highly resistant to effluents of low pH, but they are demonstrably much less resistant to oil (Perkins, 1974). Species selectivity necessitates a number of different approaches to toxicity studies. Compare the two following approaches.

1. The use of a single convenient test species to establish the relative toxicity of a wide range of known substances and of various, effluents and effluent components as a routine means of effluent quality control.

2. The use of a carefully selected range of relevant organisms from the environment to determine the toxicity of a specific polluting discharge or its component effluents.

The former is often highly stylized and may employ a species "irrelevant" in the context of the receiving waters (*e.g.* Artemia; Price *et al.*, 1974; Table 4.1) or yeast (Nelson-Smith, 1969) which may have no truly marine forms (E. O. Morris, Personal communication). Such assays have little in common with, and will make little contribution to, the prediction of field responses since they seek only to compare the toxic entities one with another. It would be foolish to forecast the likely ecological effects on such bases. On the other hand they throw useful light on biological response in relation to chemical structure; very valuable information for assessing some proposed new discharges in relation to similar existing ones may be obtained.

Where one is concerned with likely ecological effects, the most meaningful study of toxicity must depend upon the situation. What is known about responses of the community to this effluent as it affects component species ranging from unicellular algae to fish? Which species are at risk and which are the dependent species? How will effluent distribution change the distribution of species and, if there are particulate pollutants present, where are they likely to concentrate? What do we know about the relative sensitivities of the component species?

Fortunately in relation to some of these questions the literature covers a very wide range of test organisms that have been used with a varying degree of success: diatoms and macrophytic algae (Boney, 1968, 1970; Natarajan, 1970; Pybus, 1973); protozoa (Gray and Ventilla, 1971); fertilized eggs and larvae of invertebrates, including those of polychaetes, crustacea, molluses echinoderms and urochordates (Corner and Sparrow, 1956, Okubo and Okubo, 1962; Breese, 1967; Widely and Blick, 1967; Woelke, 1967; Wilson, 1968; Brown and Ahsanullah, 1971; Koybayashi, 1971, 1972; Wells; 1971; Connor, 1972a, b; Koybayashi, Nogami and Doi, 1972; Chia, 1973; Renzoni, 1973); invertebrate adults, including reef coral, polychaetes, archiannelids, crustacea, molluscs and echinoderms (Corner

TABLE 4.1

Tolerance limits for brine shrimp (from Price *et al.*, 1974)

Compound	*24 h TL_m Dosage (mg/l)*
1	*2*
Screening test, $TL_m < 100$ mg/l	
Acetic acid	42
Amyl acetate (primary)	53
Benzene	66
Diisobutyl ketone	65
Ethylacrylate	12
Ethylenediamine	14
2-ethylhexanol	19
2-ethylhexylacrylate	72
Isodecanol	3.4
Linear alkylbenzene[1]	33
Sodium lauryl sulphate[2]	3.6
Styrene	68
Tetrahydronaphthalene	78
Toluene	33
Vinyl acetate	45
Screening test, TL_m 100-1000 mg/l	
n-Butyl acetate	150
Cumene	110
Diethylenetriamine	710
Ethylene dichloride	320
Ethylene glycol monobutyl ether	1000

1	*2*
Epoxidized soybean oil	240
Formic acid	410
Isophorone	430
Isopropyl acetate	810
Methyl amyl acetate	230
Methyl amyl alcohol	370
Octyl epoxytallate	300
Phenol	160
n-Propyl acetate	820
Screening test, TL_m 1100-10,000 mg/l	
Acetone	2100
n-Butanol	2950
Diethanolamine	2800
Ethyl acetate	1590
Ethylene glycol monoethyl ether acetate	4000
Hexylene glycol	5900
Isobutanol	1400
Isobutyl acetate	1200
Methyl ethyl ketone	1950
Methyl isobutyl ketone	1230
n-Propanol	4200
Triethanolamine	5600
Screening test, TL_m >10,000 mg/l	
Di-decyl phthalates	>saturation
Diethylene glycol	—
Diethylene glycol monoethyl ether	—

(*Contd.*)

TABLE 4.1 (*Contd.*)

1	*2*
Di-2-ethyl hexyl phthalate	>saturation
Ethanol	—
Ethoxytriglycol	—
Ethylene glycol	>20 000
Ethylene glycol monoethyl ether	—
Ethylene glycol monomethyl ether	—
Isopropanol	—
Methanol	—
Propylene glycol	—
Tetraethylene glycol	—
Triethylene glycol	—

[1]11-to 13-carbon alkyl group.
[2]Included in test series as a control.
[1]TL_m not precisely determined when screening tests indicated it was >10 000 mg/litre.

and Sparrow, 1956; Raymont and Shields, 1962; Woelke, 1968; Perkins, 1968a, b, 1970, 1972a, b; Portmann; 1978; Portmann and Connor, 1968; Bellan *et al.*, 1969; Eisler, 1969; Brown and Ahsanullah, 1971; Gray and Ventilla, 1971; Lewis, 1971; Swedmark *et al.*, 1971; Wildish and Zitko, 1971; Gribbon, 1973; Perkins, Gribbon and Logan, 1973); and fish (Eisler, 1965, 1967, 1970a-c).

As a first approach, the toxicologist would normally consider the more abundant species of direct or sub-economic importance, namely resources species and organisms which stand in a trophic relationship with them. Generally, this requires studies on adult, juvenile and, where relevant, larval forms of resource species, Unfortunately, many larvae are exceptionally sensitive (Wilson, 1954) and only a few can be cultured. This is a major weakness of routine laboratory studies and of most field studies. The less abundant, though possibly ecologically important species, may be difficult to collect and acclimatize to aquarium conditions in adequate numbers for experimental work. Although the toxicologist may use large numbers of the more abundant species the cull which he exerts on them is trivial when compared with the influence of one commercial fisherman. This is not true with respect to the rarer forms on which even modest collection may have an adverse effect and thus bring about the result which it is intended to avoid. There is a very great deal to be said in favour of the general precept that all assay organisms should be cultured in the laboratory. Such a procedure permits the study and control of the whole life cycle, it does not deplete natural stocks and it overcomes the psoblems of handling and acclimation of test organisms (Reish, 1973) and ensures a uniform assay plant or animal. Unfortunately, this desirable condition can seldom be attained since many of the more important organisms have not been as yet successfully reared in the laboratory. Moreover, many of the desirable test species are relatively large and the culture facility required to maintain the necessary range of test species would represent an industry beyond the capacity of most laboratories to sustain. Consequently, laboratories must depend on the use of wild stocks for a long time to come, despite the problems of selection, handling, care and maintenance which this iniroduces (Perkins, 1972a, b).

It is widely recognized that laboratory studies have severe limitations when it comes to applying the results to the field situation and, in consequence, assessments of water quality should be of two complementary parts, namely laboratory and field models and field trials. Like laboratory assays, such models and field studies also have their difficulties and limitations.

The best investigations are so organized that the results of laboratory study, placed in the framework of the required community structure, can be interpreted in the light of the known or calculated dilutions and other parameters measured in the field. In this way, the release of effluent or toxic material can be established by experiment without recourse to an artificial margin of safety, and such forecasts can be confirmed, or otherwise, by further field investigations of the chemistry physics and biology of the situation. If necessary, further experimental work on issues less amenable to laboratory study can be conducted in the field. A good example of this is the follow-up work of the U.K. Atomic Energy Authority in relation to the selective accumulation of radionuclides in all forms of marine life in the neighbourhood of discharges. By this means any lingering doubts about the very long term effects of low level exposures can be confidently dispelled.

B. Toxicity Testing

Laboratory determinations of toxicity may be divided into short-term acute toxicity tests and long-term or chronic biological response. These are essentially bioassays based on responses on short or long application times; the former measures the more or less immediate biochemical disruption of life processes, the latter permits the day-to-day mechanisms to proceed until they become affected by the blockage or disruption of some slower metabolic processes.

1. Acute Toxicity

To an almost overwhelming degree, studies of the toxicity of materials in aqueous media are preoccupied with short-term acute exposure. As a result long lists of median tolerance limits have been compiled from the copious literature (Mawdesley-Thomas, 1971; Perkins, 1972a; E.P.A., 1973). Frequently, the conditions of the experiment are ignored or ill-defined (Perkins, 1972; E.P.A., 1973), some involve species totally irrelevant to the stated objectives of this study and, as Thom (1971) noted, some biological materials may be in such bad condition that the results reported are at best limited in application, at worst useless: often the reporting may be so vague that it is not even certain whether a 50 per cent or 100 per cent mortality is referred to and may lack references and vital experimental data. This is, however, a matter of degree. Few workers bother, or apparently consider it necessary to refer to the mode of collection, transport, handling and acclimation or other methods whereby stress may be minimized; even fewer describe the environmental conditions observed during the test which form an integral and indispensable part of the results of the study and without which it has little meanining (Perkins, 1972a, b; Table 4.2).

Acute toxicity tests are much concerned with the determination of the median lethal response or median tolerance limit (TL_m) which is no more than it says, namely the condition which produces a 50 per cent kill or permits a 50 per cent survival of the test organism. Such an approach may be applied equally to the effect of toxic gases, liquids, solutions, solid and suspensions, X-rays and such environmental parameters such as light, temperature, etc. In studies on terrestrial animals, it is possible to be sure that the dose given by inhalation, ingestion, injection or irradiation is received in total by the animal under test. Thus it is possible to talk of the median lethal dose, *i.e.*, the LD_{50}, with reference to a given animal: this then would be the dose required to kill 50 per cent of such animals so exposed under defined conditions. Since, however, time is a determinant during the exposure of an animal to an applied stress, it is essential that the duration of the exposure be expressly stipulated and this is normally 24, 48 or 96 h, in which case the result is expressed as a 24, 48 or 96 h LD_{50}, where D indicates the dose actually deposited within the animal, and the units of the dose are appropriate to the nature of the imposed stress: thus of a drug as g/kg of animal body weight, temperature as day-degrees or of radiation as rads and so on. In those acute and chronic studies where the investigator is concerned with an observable effect other than mortality, *e.g.*, anaesthesia, it is customary to stipulate the median effective dose which is the magnitude of the imposed factor required to bring about the observed effect in 50 per cent of the test animals and is indicated by the symbol ED_{50} or 24 h ED_{50}, etc.

In aquatic studies is is usual to discuss toxicity in terms of the concentration of the applied factor in the surrounding water. Here one is concerned with the median lethal concentration (LC_{50}, TL_m) or the median effective concentration (EC_{50}) which correspond to and are defined in the same terms as LD_{50} and ED_{50} respectively, but in this case do not attempt to define

TABLE 4.2

Procedures used in Toxicity Testing with Marine Invertebrates (after Perkins, 1972a)

1. Collection[1]	By hand or baited pot; when possible, avoid trawl or dredge collections	*Select*—Half grown, growing and uniformly sized individuals	*Reject*—Mishandled, deformed and non-uniform stock	*Avoid*—Making collections when the air temperature is $\leqslant 0°C$, particularly with those animals taken from below the low-water mark. Similar care will be required at the other end of the temperature scale, but such problems have not arisen at Gorelochhead
2. Transport[1]	Handle carefully; aerate	*Avoid*—Excessive exposure to direct sunlight, prolonged exposure to the atmosphere when no shelter is provided and undue disturbance or jarring. Especially avoid overcrowding		
3. Acclimation	(a) Ambient temperature experimentation. Acclimate for 2 weeks at normal tank room temperatures	Do not overcrowd; feed keep clean. Maintain salinity in the range 28 to 32 per cent. Aerate adequately. Monitor temperature	*Reject*—All stocks in which the mortality exceeds 10 per cent at this stage	*Avoid*—Maxing stock collected on different occasions

		(b) (i) Acclimate for 1 week at normal tank room temperature, then— (ii) Acclimate for 2 weeks at 20°C	and pH. Maintain an adequate record of stock	
4.	TL_m determination	Maintain salinity in the range 28 to 32 per cent. Aerate adequately. Ensure adequate replication of tests.	Dose period. Do not feed	*Select*—All individuals in normal test aliquots by random number methods. *Record*—all overt symptoms, including number anaesthetised, locomotory ability and number patently dead. If deaths in controls, reject test.
		Monitor temperature, pH and oxygen concentration	Recovery period. Do not feed	*Record*—Course of recovery from anaesthesia (if any), mortality and pattern of death; any other symptoms characteristic of the toxin
5.	Post-TL_m studies	(a) Long-term recovery from single treatment	Laboratory studies. Maintain salinity in the range 28 to 32 per cent. Aerate adequately. Monitor temperature, pH and oxygen concentration	Select a given group of treated animals, normally those from the highest concentration at which 100 per cent had survived to the end of the five day recovery period. Feed. This method is suitable for carnivores, *e.g.*, Carcinus (shore crab) and Eupagurus (hermit crab). When cannibalism occurs, compare survival in treated animals and controls, maintained in groups and in solitary confinement. *Study*—Food intake, growth and mortality. (a) Release large aliquots, including controls (more than 100 individuals per concentration) into the environment after suitable marking.

(Contd.)

TABLE 4.2 (*Contd.*)

		Study—Growth, morality and vulnerability to predation. Suitable for herbivores, such as Littorina, and carnivore, such as Thais (Nucella)
	Field studies	(b) When method (a) is unsuitable because the test animal is either too vulnerable on grounds of size or re-capture is likely to be a problem, treated animals and controls can be cultured in live boxes suspended from a trot line. Subsequently, larger individuals can be cultured by normal methods. Suitable for Buccinum (whelk) (not re-captured easily); Ostrea (oyster) and Chlamys (queen scallop) (too vulnerable at a small size). *Study*—Growth, mortality, reproductive status and taste
(b) Long-term chronic studies. Maintain salinity the range 28 to 32 per cent. Aerate adequately. Monitor temperature, pH and oxygen concentration	Expose small numbers (5-8) of test animals to 40-l volumes of toxin solution at concentrations that are low or very low compared with the TL_m. *Study*—Feeding, growth, development of mortatity and any other parameter pertinent to the particular animal and toxin. Feed test stock and controls on the day prior to the renewal of the test dose. At the time of the renewal, wash the tanks thoroughly with fresh water; scrab when necessary. Scrupulous altention to tank hygiene is essential, Stock must be collected, handled and acclimated according to the criteria 1-3 above. Death in controls should be regarded as undesirable. Strong justification is for the continuation of an experiment once control death develops; such justification may include cannibalism at the time of ecdysis, but must be used with care.	

the dose actually deposited within the animal. This arises because unlike studies on terrestrial organisms for which it can be certain that the test material has been ingested, often the only thing certain in aquatic studies is the initial concentration which was introduced into the water. Except for studies on the uptake of radioactive nuclides, there have been few attempts to establish the dose actually experienced inside the plant or animal. In general, by the standards of competent cage animal experiments, most aquatic bioassays are crude. It is relatively straightforward to measure the intake of food and loss of urine, faeces and respiratory requirement of a land animal. Compare this with the difficulties of working in the all-pervasive environment of the aquatic animal. Water is its support medium, bringer of oxygen, repository for excretory and respiratory products, vehicle for food and source of many chemical stimuli. The animal conditions its environment in many complex ways including osmotic and ionic effects as it lives. Some fortunate species have the facility to seal out the medium and thus survive transitory adverse conditions.

Even the study of radioactive nuclides is much more complex in the aquatic environment than it is for animals on land where the fate of ingested foods and its progress during assimilation and excretion can be measured with relative ease (Bailey and Bunyan, 1972). In land animals, it is possible to make detailed investigations of particular aspects of body function with only moderate difficulty. Aquatic animals do not allow easy measurement of separate body functions, for example osmotic and ionic regulation depend on the medium for their operation, from which ions are withdrawn and to which excess ions are returned along with excretory products, so that exchange after a brief initial period becomes confused. For this reason the most successful experiments on marine animals in sea water rely on labelled doses introduced in food or by injection into the animal.

Some toxic materials, such as acids (Doudoroff and Katz, 1950) and detergents (Marchetti, 1965), act primarily on the external epithelia. In such situations, the matter of whether such application of material should be regarded as LD_{50} or

LC_{50} is polemical. In the event, the results obtnined in aquatic studies are best understood when they are expressed in terms of concentration (*i.e.*, ppb; ppm; ppt) as a 24, 48 or 96 h LC_{50} (or EC_{50}) in Britain and as a 24, 48 or 96 h TL_m in North America. Use of these terms is clearly something of a matter of personal preference, though the North American convention does have some advantages when used with those parameters that cannot be expressed in terms of concentration, *e.g.*, temperature. For more detailed information on synonymy of terms, reference should be made to Sprague (1969) and Batelle Columbus Laboratories (1971).

Thus the use of EC_{50} (environmental concentration) or equivalent term offers the only practical solution to the problem of expressing the exposure of an organism to a pollutant, at least for the present in most cases. This practical convention does, however, represent a very big gap in a large part of existing knowledge of aquatic toxicology.

Thousands of millions of dollars are being spent on the construction of facilities designed to stem the flow of pollutants into our continental waters. Tens of millions are being spent on research aimed at matching existing technical knowledge with these national needs. Almost nothing is being spent on research devoted to the ecosystems we seek to save (after W.S. Brocker, Science 182 (4111), Editorial, p. 435, 1973).

Without this essential ecological knowledge, the value of the currently fashionable and expensive mammoth programmes of "monitoring" of residues such as DDTs, PCBs and the like in all manner of marine biota must be questioned, when so little effort is made to study the effects in the round. If marine science is truly concerned with the maintenance of healthy and productive stocks of resource animals and their associated foodwebs then the answers must be sought very far from the present locus of research tied to short-term toxicity criteria and the ingestion of "total pollutant" in the food of man. If we must take an anthropocentric viewpoint then it is futile while pursuing the goal of preventing the cumulative poisoning of man to ignore the fact that in turn his food organisms must also be of high quality and healthy when they are alive.

With this fundamental caution, short-term toxicity testing remains a useful tool, provided it is understood what this piece of information means and what the limits of its valid use are.

In all cases, whether one is concerned with the determination of the LD_{50} or of the LC_{50}, the course of the experiment follows the same basic procedure. A preliminary experiment is undertaken to determine approximately the order of the effectiveness of the applied stress. Then aliquots of test animals are placed in a range of test conditions, which may or may not be arranged in a geometric progression, and are exposed to these conditions for the stipulated time period. At the end of this period, the number of deaths in each concentration is recorded. In marine studies, the results obtained are normally based on the survival of the test organism maintained in fresh, clean sea water for five or seven days after the stipulated duration of exposure (24, 48 or 96 h say) to the toxic entity. This procedure allows anaesthetized individuals to recover or any consequential mortality to develop (Perkins, 1972a). However, at every stage from the initial collection of the test animals to the end of the recovery period great care and built-in checks against the influence of unhealthy, poorly acclimated stock and of the fuilure of the test equipment are necessary (see Table 4.2.) The results recorded at the end of the recovery period may be analysed by a variety of statistical methods which include graphical interpolation and probit analysis (Sprague, 1969) and structures of great statistical complexity may be based upon such simple measurements. Leaving aside for the moment the manner in which such data is ultimately used, it seems incredible that such methods are in current use and yet Sprague (1969) in his excellent review was forced to point out that, even for freshwater fish, remarkable little is known of the requirements for acclimation before a test and of acclimation during the test.

With respect to marine animals, the situation is rather more adverse since in general so little is known about their maintenance for prolonged periods in the laboratory. Here it seems very clear that whatever animal or animals may be the chosen vehicle of a laboratory's work, that laboratory will first have to determine the more essential needs of these animals at all stages of their life history. For example Ross (1974), who worked on the sand goby Pomatoschistus minutus, found that it was not just the volume of water available to each fish, but a minimum area of tank bottom, about 100 cm^2/individual, which was essential to its continued health and survival. Similarly, O.J. Abbott, who worked in my laborotory, found that in the absence of sediment the cockle Cerastoderma edule shows a very poor survival. When, however, there is a sufficient depth of sediment for these animals to submerge totally, the mortality is very low, even when the superficial water is reduced to a depth of only 2 cm. In the absence of such understanding, the use of elegant statistical techniques does seem to be rather wasted.

Second, the detemination of the TL*m* is often performed in a highly stylized manner (Doudoroff *et al.*, 1951; Lennon, 1964; American Public Health Association, 1965; Portmann and Connor, 1968; Cairns, 1969; Cairns. Waller and Smrchek, 1969; Sprague, 1969, 1973; Portmann, 1970; Perkins, 1970, 1972a; Connor, 1972; Howard and Walden, 1972; Saward, 1972; Wilson, 1972), and few attempt to define any of the behavioural and clinical symptoms that may be evinced during the investigation (*e.g.* Lennon, 1964; Perkins, 1970, 1972a; Saward, 1972). Some laboratories allow the test organisms a recovery period of five or seven days; others dispense with this part of the procedure completely, thus the technique provides no information regarding the onset and scale of delayed mortality which may not be apparent in the initial exposure period, but which may go on for a long time once it has started (Wilson; 1968; Perkins, 1970, 1972a). Because it is so stylized, the applicability of this test to environmental problems is limited; indeed it is equally uninformative regarding the effects of very short-term and truly long-term exposures. As Warner (1964) said

> Death is, by definition, the cessation of biological activities. Hence it is asymptomatic; and no information relevant to sub lethal effects can be obtained from the response

and Fry (1971) stated that

> The pollution biologist, on the other hand, wishes to protect

> the organism from damage not from death alone...Lethal limits are to be considered only as boundaries within which the real work goes on.

Third, as Holden (1954) noted, a 50 per cent mortality is not a margin acceptable to fishery interests, it is, therefore, usual to ascribe a notional "application factor" to the TL_m (usually a 96 h LC_{50}) and hopefully assume that all will be well in the environment. Such application factors may be based on information other than the median tolerance limit (F.W.P.C.A., 1968), but in general and to quote Sprague (1971)

> The application factor, as used in fisheries and pollution work, provides a safe level which is not known from a median lethal concentration which is known. The LC_{50} is multiplied by an application factor, such as 0.1, to obtain a concentration which presumably has no sub-lethal or chronic effects. The value of the application factor is assigned on the basis of judgement of scientists. Usually this judgement takes into account any evidence available on the relation between safe levels and lethal levels. However, in the absence of such research knowledge, as is frequently the case at present, the judgement may have to be rather arbitrary.

In effect then, and in practice generally, the application factors used represent a compromise which may be some contrived extrapolation of LC_{50} results or some rule-of-thumb empirical multiplier applied to it. Such factors have no sound technical basis (Howard and Walden, 1969), nor is there any rule that physiological, biochemical or epidemiological processes operate in this manner. They can be misleading, if not mischievous, when applied to pollutants such as detergents whose dose-response curve is highly skew or even paradoxical (Schatz *et al.*, 1964; Perkins, 1970).

Manifestly, neither the use of confidence limits nor any amount of mathematical analysis, however elegant, can ameliorate this fundamental weakness, since short-term lethal toxicity tests can do no more than indicate the order of magnitude of the toxicity of the material in question. Likewise, the relative merits of the 96 h TL_m as compared to the 24 h TL_m and flowing or static experiments must be regarded in the light of these considerations as a matter of little consequence, since none can provide a valid basis on which to formulate recommendations for determining concentrations acceptable in the environment. True, the situation may be improved somewhat by the construction of toxicity curves based on a series of TL_m, thus 0.5, 1, 3, 6, 12, 24, ... n h, but since the most prolonged exposures seem to be 30 days (Sprague, 1969) and the simple death/survival observation remains the critierion of toxicity, the basic position continues to be scientifically untenable.

What then are the positive virtues of the median tolerance limit? Regrettably, they are few. The conditions under which TL_m can be used validly in aquatic toxicology must be clearly understood.

1. Fry (1971) has demonstrated that toxicity tests set the boundaries within which may be performed the really close studies of short term subacute effects and long term studies of chronic toxicity, physiological effect, effects upon reproduction in relation to the whole life history and the study of sub-lethal response in general.

2. It is just not possible to undertake detailed studies on every species living in a given situation. Nevertheless, it is often very important to have some assessment of the comparative toxicity of ranges of pollutants to a much wider variety of organisms than those that can be subjected to as detailed a study as suggested above.

   The expression derived by Dowden (1961), viz.

$$\text{Cumulative Toxicity Ratio} = \frac{24 \text{ h } TL_m}{48 \text{ h } TL_m}$$

   is a rule of thumb means of measuring the way longer term toxicity testing of a pollutant might go and may be used to advantage in planning further experimentation.

Here then the median tolerance limit has a valuable role to play, particularly in directing ancillary studies of the behaviour and response to a given pollutant over a range of time intervals.

3. Once the relationships that a TL_m of whatever duration bears to the considerations laid out in (1) and (2) are thoroughly understood, then, and only then, may the median tolerance limit be used in what is potentially its most valuable role, namely the routine screening of a known pollutant or given effluent as a measure of quality control. This application is of special practical value when complex industrial effluents must be monitored and adequate chemical analysis may be extremely demanding and in the event may only be able to quantify some of the toxic components, whereas the bioassay gives an immediate evaluation of the total system (Wilson, 1972). All scientific work is costly, even the simple TL_m study, but by using a test organism carefully chosen in the light of (1) and (2) a simple routine bioassay based on perhaps two or three test aliquots may well provide a cheaper, more efficient and more positive measure of effluent quality than the more pretentious and possibly more demanding and more expensive methods of chemical analysis. Moreover, in adverse conditions, when chemical quality control fails or has doubtful validity, the apt biological assay is relatively inexpensive, fairly rapid and usually unambiguous. It is of interest that it is on this basis that B.C. Research, Vancouver, provides an effluent control service to the wood pulp industry of that province. This service is based on a specified assay procedure and a standard for the 96 h TL_m of pulp-mill effluents laid down by the Provincial Government. Each week, every pulp mill in the province sends an appropriate representative effluent sample to B.C. Research, who determine the 96 h TL_m with respect to a stipulated salmonid fish.

As a corollary, it should be noted that for a regulatory body faced with the problem of monitoring possibly hundreds of effluent releases this method may provide (a) the only simple low-cost method of keeping a general watch over its sphere of operations bearing in mind the limitations of the method and (b) information on the comparative acute toxicity of materials at various stages of the process. Therefore, it can be used to identify and focus effort upon those component effluents or polluting sources most in need of immediate, more detailed attention. In this way it is often possible to reduce or eliminate product or reagent losses thus effecting a useful saving. Or, if effluent treatment is needed, it may be that only certain streams need be processed thus saving in capital outlay and running cost.

4. When it can be so arranged the duration of the exposure to the toxic entity in the laboratory should be equal to that which takes place in the environment. This is of a particular relevance in the determination of the toxicity of solvent emulsifiers used in the clean-up of oil spills. Here the core problems are concerned with the paradoxical dose response and the normal maximum duration of the acute exposure is less than six hours. Clearly the test of acute toxicity ought to be framed either in terms of a 6 h LC_{50} or the 1 h extreme acute exposure used by Perkins (1968a, b, 1970) and Crapp (1970). Regrettably some regulatory and dispersant-testing bodies apparently insist on either 48 or 96 h test exposures irrespective of whether this is realistic in relation to its use in the field or not.

Regarding the Torrey Canyon accident, Smith (1968) showed that with few exceptions such as high level rock pools, dilution occurred rapidly on the open shore and only very low concentrations were recorded in the inshore waters. Clearly, 48 or 96 h tests are not relevant here, since they give no appreciation of the peak risk or lack of risk in the period of about six hours during which time the situation changes from most acute to negligible. From another viewpoint, we know that even if lateral water movement is minimal, tidal movement will ensure that any oil emulsifier will be transported away more or less rapidly from the site of application to affect another area in diluted form. The field situation could be more nearly represented by an acute exposure of the test animals to a range of high concentrations of oil plus dispersant, or dispersant alone for a period of six hours. At the end of this time, the test solutions should be progressively diluted to very low concentrations, without change of vessel and the test organisms exposed at effectively clinical concentrations for a period equivalent to that taken for the toxicants either to be transported away from the area and replaced by clean sea water, or where residual transport is inadequate until total biodegradation had occurred.

The problem of toxicity assessment in relation to dispersants as they are applied in an intertidal situation is one example of how practical tests can be made to imitate important features of the environmental situation. Here is another. It is a basic assumption of routine effluent toxicity testing that a stable pollutant is uniformly discharged into receiving water of constant "assimilative capacity". Instead of this uniformity, variability is the familiar experience of those daily involved with the practical problems of effluent control on spot samples. It is necessary to select typical samples of the component effluents and perform TL_m tests over a range of exposure times in water samples taken at representative stages of the tide. Having established the most critical combination of effluent composition, exposure time and receiving water, it becomes possible to curtail the range of tests needed. There are many other possible examples of foreseeable incidents such as accidental release or plant failure which can be thus taken into account.

In my opinion, while some authors like Wilson (1972) do concern themselves with the dose response at high concentrations, the determination of toxicity tends to become a ritual based upon a standard 48 or 96 h TL_m. Such an approach clearly ignores the effects of dosage at high or very high concentrations, both in terms of direct mortality and of subsequent ecological effects. In terms of the latter, at least, it would seem to be as desirable to extend the study of toxicity into the field of extreme acute toxicity as to undertake long-term chronic studies.

To summarize, the estimation of the median tolerance limit must be considered a minor tool in the hands of the aquatic toxicologist. It is particularly susceptible to misuse but it can, when its relationship to the broader aspects of toxicology are properly understood and defined, become a valuable tool in the hands of those concerned with the control of effluent quality. Use outside these well-defined limits is invalid. Despite the many inadequacies just discussed, it is very clearly better to have properly conducted acute toxicity studies rather than sundry chemical analyses which may have no bearing on the assessment of effluent quality.

2. *Sub-Acute Biological Response*

There seems to be some confusion in the literature regarding the definition of a long-term sub-lethal study. Long-term may refer equally to an experiment of perhaps 240 h or to one lasting a year or longer. Analysed objectively, an experiment of 240 h duration might be considered as long-term if it referred to micro-organisms or other components of the biota having a life cycle of a few hours or at longest a period of two weeks. For most normal test animals (shellfish, fish, etc.) a period of 10 days is a small proportion of the lifetime and the experiment should not be described as long-term.

Equally, there is very considerable confusion about what constitutes a sub-lethal dose of a toxicant, or indeed what is meant by the term sub-lethal; even Sprague (1971), in an otherwise excellent review, does not make this clear. Probably the clearest thoughts on the subject are those of Alderdice (1967), who defined sub-lethal effects of pollutants as debilitating and thus only bringing about death indirectly. Studies concerned with the clinical toxicology of such responses are likely to be important in relation to fishery productivity. In practice, however, and regrettably in most cases, a "sub-lethal" dose is taken to mean one that is a notional fraction of the TL_m, assumed to be safe, but which in the long term may prove fatal to all the exposed organisms (Table 4.3). Clearly, such a dose, is not sub-lethal and the experiment should fall into the sub-acute category.

Not doubt part of the problem arises from the character of the behavioural and clinical symptoms that are capable of assessment consequent to sub-acute or long-term exposure. In both conditions, and particularly at levels of clinical dosage, toxicants may be expected to induce one or more of a number of relevant and important effects which include:

1. inhibition or enhancement of growth,
2. influence on reproductive potential resulting in changes in fecundity and egg, larval or brood survival,
3. behavioural changes such as the influence of toxicants on the settling ability of the planktonic larvae of shell-fish or altering the responses of larvae and adults to environmental factors.
4. an upsetting of predator-prey relationships,
5. an inhibition or modification of the crustacean moult cycle,
6. effects on respiration, osmo-regulation, ion-regulation, and other biochemical and physiological mechanisms,
7. predisposition to parasites, disease and histo-pathological changes.

However asymptomatic mortality may be, it is unambiguous, whereas the relationship between the observation of a particular clinical symptom and the prediction of an undesirable effect on a popoplation may be difficult to establish (Mount and Stephan 1967).

(*a*) *Intermittent exposure*: The study of sub-acute toxicity is especially important with respect to those species which may come within the ambit of an effluent or any non-lethal toxic

TABLE 4.3

Comparison of Median Tolerance Limits and Long-Term Chronic Exposure of Marine Invertebrates to Toxic Solutions (all salinities=30‰) (After Perkins, 1974)

Species	*Substance*	*24 h LC_{50}*	*Chronic exposure*		
			Concentration	*Fraction of LC_{50}*	*Time to 50% mortality*
Eupagurus bernhardus	Na_2SO_4	> 6400 ppm	500 ppm	< 1/13	12 weeks
Eupagurus bernhardus	Board mill effluent	> 50%	1%	< 1/50	43 days
Eupagurus bernhardus	Gas flume	> 10%	0.5%	< 1/20	17 weeks
Carcinus maenas	Na_2SO_4	> 6400 ppm	275 ppm	< 1/23	> 8 weeks
Buccinum undatum	CH_3COOH	⌐ 6400 ppm	100 ppm	⌐ 1/64	12 weeks

[1]96 h LC_{50}

event for a relatively short period, but which are not subjected to a lifetime of chronic exposure to a toxic entity. Clearly the diadromous fish such as the eel and salmon which may have to pass a barrier of low-quality water during their migrations to and from fresh water fall into this category. Equally, this type of assay is of importance to those fish and vagile epifaunal species which may encounter an outfall or dumping during their migrations or wanderings, but which can, should they survive, move into cleaner water.

(*b*) *Avoidance*: Structural damage, *e.g.*, erosion of taste buds (Bardach *et al.*, 1965; Forster *et al.*, 1966); changes in conditioned response (Warner *et al.*, 1966) and avoidance reactions (Sprague and Drury, 1969), have all been noted in freshwater fish. Here avoidance behaviour may be particularly important, since such a reaction prevents the organism from becoming exposed to the worst effects of the pollutant. It is of no ultimate benefit if avoidance of a polluted area leads the fish into another river where it cannot fulfil its migratory mission successfully. However, as Sprague (1971) noted, there is no generalized description of typical avoidance reactions and, with the apparent exception of sharks (Tester, 1963) this work has been confined to salmonids or freshwater species

Where contact with an effluent cannot be avoided, as for example where a diadromous fish must swim upstream and in so doing pass a barrier of low-quality water, then it is important to assess its ability to pass this barrier. The most satisfactory test is that of swimming performance. According to Sprague (1971), rather few of these studies have been undertaken and with somewhat variable results, again with reference only to freshwater fish. However, Howard (1973) was able to show that the swimming stamina of coho salmon, Oncorhynchus kisutch, was affected by a kraft-mill effiuent concentration between 0.1 and 0.2×96 h TL_m. Furthermore, he was able to show that the temperature tolerance of rainbow trout, salmo, gairdneri, was affected by effluent concentrations ranging from 0.062 to 0.230×96 h TL_m. Whereas, the procedure used in the swimming tests is critical to its success, the temperature tolerance test which he developed is less susceptible to operational problems and may be used as a practical routine bioassay to measure sub-acute "safe" concentrations of effluent for discharge to receiving waters.

(*c*) *Respiration, photosynthesis and growth*: The respiratory response may be important to survival in sub-acute conditions. The frequency of "coughing" is fish, described by Schaumburg *et al*,. (1967) and thought to be a useful measure of response, changes in the presence of sub-acute concentrations of kraft mill effluent but is, according to Howard (1973), difficult to quantify; this work too is concerned with salmonids and in the freshwater context only. Regarding marine creatures, it has been found that the rate of respiration of the lobster is directly related to the O_2 tension (Thomas, 1954); a response which is an important aid to survival in adverse conditions (Perkins, Unpublished work). In other cases the interpretation of the respiratory response is less certain.

It was shown by Hochachka and Teal (1964) that the respiration rate of Gymnodinium nelsoni could be influenced by a variety of inhibitors such as arsenite, fluoroacetate, ascorbate. In general, however, it, seems fair to state that there is a consensus of opinion among workers that studies of respiration are likely to be unrewarding. The photosynthetic rate of unicellular algae may also be inhibited and can be used for sub-acute tests (Sprague, 1971). For example the Marine Laboratory at Aberdeen use Phaeodactylum tricornutum or Skeletonema costatum as a bioassay procedure along with other tests for the appraisal of complex effluents in marine situations where there is eutrophication. To facilitate measurements radioactive $^{14}CO_2$ is added and the uptake of labelled CO_2 provides a ready means of accurately determining the photosynthetic rate. If desired, respiration measurements can be made in like manner using a "dark bottle" technique.

Exposure to a single acute does of a pollutant may kill some members of the species, but the survivors may also display adverse effects at a later date even if no further mortality occurs. Of these adverse effects, growth is clearly an important parameter with respect to the potential gross yield from a fishery; and effects on growth have been studied particularly in relation to oil and the oil emulsifiers (Bryan, 1969; Perkins, 1970; see also below under field studies).

(*d*) *The prey predatory relationship*: The prey-predator relationship is obviously of importance, and Perkins (1970) has shown that the relationship between the shore crab, Carcinus maenas, and the rough winkle, Littorina saxatilis, was changed by an acute, but apparently non lethal does of BP 1002 administered to the latter. Later Gribbon (1973) showed that the dogwhelk Thais (Nucella) lapillus, was more susceptible to predation by the whelk, Buccinum undatum, subsequent to treatment of the former by detergent. However, the results obtained may not be so unequivocal, for Gribbon found that when treated oysters Ostrea, were offered to the predators Carcinus and the harmit crab, Eupagurus, the controls rather than the treated animals were selected; the essential problem here is that Ostrea is not a normal component in the source environment of the particular predators used in the experiment, moreover, the oysters once treated could have been more distasteful to these animals than were the controls.

(*e*) *Behavioural effects*: The scope for the study of altered behaviour in response to pollutants or imposed chemical stimuli is vast. Most if not all marine animals are equipped with chemical sense organs that fulfil a great many useful functions. In some, sense organs assist feeding or food selection, in others chemical stimuli are involved in schooling behaviour, recognition of the same or opposite sex and in reproduction, to mention only a few. Although almost any animal could be used, the range of responses is so numerous and for many species behaviour so complex when studied in detail that choice must be narrowed down to those animals with very limited "forms of expression" and/or that display a pronounced and measurable behavioural activity.

Behavioural response assays are more meaningful for fairly well specified pollutants (such as oil or pulp-mill effluents) or better still defined substances, otherwise it will be unclear what is being measured since mixed pollutants may include more than one detectable effect which may well prove confusing.

The sensitivity of behavioural assays is very great. For example the short-term acute toxicity of crude oils is of the order of 10^4-10^5 ppm for fish 10^3-10^4 ppm for crustaceaus plankton and 10^2-10^3 ppm for larvae. These values related to laboratory conditions where the test organisms are in a vessel with only a few centimetres of water beneath the oil film.

In contrast, the behavioural responses may be in the 10—100 ppm range notably, lobster—searching, feeding, grooming; crabs—"mating" stance; oysters—feeding rhythms; barnacle larvae—phototactic response. Bullheads (small fish) have a complex social behaviour and are extremely sensitive to crude oil which interferes with their highly developed olfactory senses (M.I.T., 1973).

Some of the most interesting and potentially useful assays for specific polluting situations arise from intelligent observation. In the routine assessment of 96 h LC_{50} of phenol, Stirling (1970) observed altered behaviour in the test animal Tellina tenuis. This bivalve mollusc lives buried in the sand and feeds at the send-water interfaces by means of a long siphon. When removed from the sand, they more or less immediately begin to burrow to escape predators and wave action. When phenol is present the burrowing rate is greatly decreased. It is a simple matter to expose a number of Tellina above the sand in a dish and determine how quickly they burrow. The 96 h LC_{50} for Tellina and phenol lies between 700 and 800 ppm at 9°C. From a study of burrowing rate, it was found that 10 ppm phenol had a significant effect and concentrations above 50 ppm more or less completely stopped the burrowing response while the pollutant was present. Responses were also studied in clean water over sand that had been exposed to phenol and in solutions in the absence of sand to study the closing of the shells. In exactly the same way, Tellina is sensitive to Cu in solution and to Cu-treated sand. The underlying objective of these tests was to see how pollutants affected the availability of Tellina siphons as a major food source for baby plaice growing up in the shallow tidal and subtidal waters over sandy beaches. Certain plaice nursery areas are now threatened by phenol or Cu as contaminants.

(*f*) *Disease*: Perhaps the area of work least examined and most in need of investigation is that of the relationship between the susceptibility to disease subsequent to treatment by single dose or chronic exposure to toxic substances. Pippy and Hare (1969) reported that an epidemic of lethal ulceration in salmonid and other fish caused by Aeromonas liquefaciens was probably potentiated by a combination of high temperature and pollution by Cu and Zn. Friend and Trainer (1970) showed that a single sub-lethal dose of PCBs predisposed the mallard Anas platyrhycos to infection, and subsequent death from Duck Hepatitis Virus. Sprague (1971) regarded the evidence on this topic as equivocal, nevertheless there are, for example, an increasing number of reports of an association between polluting sources and epidermal lesions in fish (*e.g.*, Wellings *et al.*, 1965; Halstead, 1970; Mearns, 1973) and others in which an association between disease and polluting source has not yet been, but may eventually be, excluded (*e.g.* Perkins *et al.*, 1972; Mearns, 1973).

(*g*) *Industrial toxicology*: Perhaps the most thoroughly studied industrial effluents are the pulp-mill wastes which have been investigated with respect to a wide range of clinical effects (Table 4.4). As with all clinical tests, these studies are ultimately concerned with the performance of fish under stress, and in particular when a given stress is applied, its influence on the ability of the fish to fulfil its normal activities. Some rivers are highly important both to the wood pulp industry and to fishing interests, for example the Fraser River. When migrating to the headwaters to spawn, the salmonid fish must pass through water contaminated by the effluents of one or a series of pulp mills. In this context there are clearly two essential questions to be answered:

(a) does the "odour" of the component materials of the effluent prevent the passage of these fish upstream and
(b) if (a) is unimportant, what influence does the presence of these materials have on fish that are under physiological stress and must make considerable physical effort in order to reach the spawning grounds and reproduce successfully?

TABLE 4.4

Responses used in Detecting and Measuring the Pollutive Effects of Pulp-mill Wastes to Fishes (after Alderdice, 1967)

Nature of response	*Ultimate effect on organism*	*Laboratory tests*
Lethal or acute		
General and gross symptoms of distress	Death	Gross determinations of biological activity and relative potency of pollutants (*i.e.*, TL_m studies)
Sub-letha or chronic		
Behavioural	Preference or avoidance reactions	Behavioural studies using "choice" situations
Physiological and histopathological	Changes in:	
	oxygen uptake	Respirometry
	enzyme activity	Dehydrogenase, transaminase acetylcholinesterase reactions
	haematological	Red cell counts, defferential cell counts, Haematocrits, haemoglobin
	serum proteins	Electrophoretic studies
	decrease in glycogen, RNA, changes in bile duct, portal vein	Histological and cytochemical examination of: liver
	decrease in RNA, change in structure	pancreas
	necrosis of tubules	kidney
	changes in secretion of mucus, activity of goblet cells, necrosis of epithelium	intestine
	changes in secretion of mucus, necrosis of filaments	gills

In these conditions, it becomes important to understand the influence of pulp-mill effluent on the ability of the fish to swim upstream and the changes in body physiology and biochemistry which accompany relatively short but intense bursts of swimming activity both in normal conditions and in the presence of pulp-mill effluent at sub-acute concentrations. For these very practical reasons, much effort has been expended on relatively short-term studies of salmonids in presence of pulp-mill effluents. That this work has been awarded such priority stresses the importance of both sources of livelihood to the provinces and states concerned. By this approach, a reasonably clear understanding of the significant interactions is emerging and a rational approach to management of the whole resource is becoming possible.

However, in relation to both this and other polluting sources, it does seem that most workers are concerned to treat the various behavioural and clinical aspects of intoxication at the acute or sub-acute level only, though the post-treatment study is prolonged in some cases (*e.g.*, Tomita, 1934; Bouck and Ball, 1965, 1966; Kramer and Smith, 1965; Smith *et al.*, 1965; Smith *et al.*, 1966; Howard and Walden, 1967; Schaumburg *et al.*, 1967; Gibson *et al.*, 1969; Butler *et al.*, 1970; Costell *et al.*, 1970; Davis, 1970; Engel *et al.*, 1970; Ernst, 1970; Mitrovic, 1970; North *et al.*, 1970; Pickering and Thatcher, 1970; Smith and Oseid, 1970; Walden *et al.*, 1970; Christensen, 1971-1972; Burton (*et al.*, 1972; Sparks *et al.*, 1972; Rosenberg, 1972; Sparks *et al.*, 1972; McLeay; 1973; Walden and Howard, 1973); moreover, as Marchetti (1965) pointed out with respect to detergents, there are few studies of synergistic effects and so far as I am aware none published concerning the effect of effluents derived from different sources, but which mix upon release to the environment. However, Eisler (1970) noted that it had become the practice among industrial toxicologists, physicians and others concerned with the health of persons who are habitually associated with dangerous chemicals to examine them by means of stress profiles. This approach is based on the reasonable hypothesis that metabolic disturbances are more likely to be discovered if a number of parameters are considered simultaneously and the results plotted on a stress profile.

This seems to be a technique that holds great promise. Nevertheless, this and many of the examples quoted above must suffer, to some degree, from the difficulty pinpointed by Mount and Stephan (1967) that the relationship between a particular clinical symptom or set of symptoms and one or more undesirable aspects of the environment of a population may be obscure; it may, therefore, be difficult to use the clinical symptom whether derived from an acute, sub-acute or long-term exposure to identify a toxic entity.

II. THE ASSESSMENT OF LONG-TERM BIOLOGICAL RESPONSE

When performed protracted chronic studies like the acute studies have by and large been preoccupied with mortality after a long-term exposure to a low dosage of a toxin (*e.g.*, Perkins, 1972c). Mortality may be asymptomatic, but when it happens the researcher or water-quality administrator is in no doubt of the direct interpretation. Accepting that clinical studies as a whole may be difficult to interpret, it is nevertheless true that valuable information, can be derived regarding residues acquired by man through his food and equally from long-term studies of the excretion rate of specific toxins from potential food organisms and from sea foods (*e.g.*, Jarvenpaa *et al.*, 1970; Tillander and Miettinen, 1970). Of far greater potential practical value and immediate benefit in toxicity assessment are those studies concerned with the interference of toxins either with the trophic relationships such as food intake, efficiency of conversion and growth (*e.g.*, Cairns, 1966; Brown and Ahsanullah, 1971), or with the ability to reproduce successfully (*e.g.*, Arthur and Eaton, 1971; Wildish, 1972). Other investigations of relevance to this problem have been carried out in such a way that they may more properly be called field trials and are dealt with as such below. These are just a beginning to the enumeration of what may be. termed long-term toxicity assessments which as will be shown can take many shapes. One approach to classification of these assessments is given below; it is probably not fully comprehensive.

A. Response relating directly or through an established link to man.

1. Monitoring of the environment.
2. Input to man.
3. Food chain studies.
4. Laboratory evaluations including biochemical pathways.

B. Response mainly involving the success of marine species.

1. Feeding studies, food conversion and growth.
2. Reproduction (including egg survival, embryonic development, larval metamorphosis, and at a later stage maturation, fecundity, fertility, and behaviour aspects of sex).
3. Marking experiments and live boxes.

C. Field Surveys and other studies designed to evaluate ecological interactions.

1. On site application of pollutants.
2. Field observations in relation to existing discharges.
3. Pilot scale models of simple ecological groups.

This classification, like any other dealing with broad and interrelating aspects of response to stimuli or to change in the environment, includes some aspects such as food or feeding that may come under two or more headings and touches on techniques that might equally be applied under various sub-headings. Of necessity, mention is made only once of each aspect or technique but this does not imply that these lack wider interpretation in relation to other sub-headings or experimental situations.

A. Response Relating to Man

In relating response to a pollutant in the environment to the eventual passing on of a certain amount of pollutant to man, a gross integration of a special kind is made which cuts across many pathways for the pollutant in the food web. Man is basically selfish and fearful of his future. He is immediately concerned with any threat or harm that may arise, including any dubiety about the wholesomeness of his food. In recent years, man's response to real or imaginary effect of pollutants in his food, including sea food, can be regarded incontestably as a form of bioassay. Having made this point, it is beyond the scope of this chapter to refer to the forms of response in detail; in principle it is reflected in his concern about the marine environment, about input to man and about the biological and biochemical processes contributing in some way to the complex pathways of pollutants to man.

1. Monitoring the Environment

The main objective was to follow the progress of pollutants, especially certain metals and organochlorines, along pathways involving the air, fresh water and surface and subsurface sea waters to sediments and within marine life at different trophic levels. In relation to biological response, the results from monitoring programmes can be regarded not just as chemical pathways or as an insult to man's sea food but also as biochemical processes in marine life. Equally, monitoring results are an indication of water quality. In essence, monitoring of the sea and marine life is a watching brief to be maintained for an indefinite period to ensure that man does not excessively pollute the seas; commercial loss of marine resources and damage to the supporting foodweb are less immediate objectives. Also in an oblique way every valid bioassay relating to existing pollution is potentially monitoring a very small segment of what is or what may approximate to the marine environment.

2. Input to Man

In a very tragic way, the disasters to man from Hg and Cd pollution (pp. 193-195) are extreme examples of the World-wide response to pollution of the sea that is going on year after year at an insidious and increasing rate but nevertheless undetectable by most chemical criteria. Local accumulation generates high spots but metals and persistent organic chemicals can also build up in man from marine products taken at some distance from any input. Only in a very few extreme cases can one regard man's death or ailment as an ultimate bioassay of a marine pollutant. In the vast majority of cases man acquires pollutants from a large number of sources of which marine food may rank as a small or insignificant contributor. Human toxicology is a topic beyond the scope of this article; suffice it to point out that for example in relation to standards set for drinking water, national and international regulatory bodies have declared limits for toxic elements at very low concentrations little different from concentrations found in clean water supplies. Pollution from Pb, As and Hg have had an enthralling history. In particular, the recent Hg scare and the problems underlying it are especially relevant to marine biological response and toxicity, although obvious damage has been confined to sea birds and the higher predators

3. *Food Chain Studies*

Very few studies indeed have been aimed at the controlled introduction of toxic elements or excessive nutrients on natural food chains terminating in commercial food species. Much more work has been done on the feeding of "laboratory" caltures of unicellular algae to small animals and on feeding "laboratory" animals such as Artemia to carnivores in a wide range of studies including larval survival and feeding. In some of these studies, the fate of introduced toxic metals and other persistent pollutants has been determined.

There is little comparable work in the marine field to that initiated by scientists at the Marine Laboratory, Aberdeen. Two contrasting food chains are being studied, the benthos-linked chain, phytoplankton-zooplankton-Tellina tenius-O-group plaice (Pleuronectes platssa) which requires a sand substratum;

and more recently the zooplankton-linked chain, phytoplankton-zooplankton-sandeel (Ammodytes). In essence, the benthos-linked food chain is a model of that for demarsal fish, especially bottom dwellers feeding exclusively on benthic invertebrates. The published results relate only to this food chain which has been used in relation to added nutrients (Trevallion *et al.*, 1973; Johnston, 1973) and a succession of toxic metals, Cu (Saward *et al.*, 1972), Hg (Saward *et al.*, 1972) metal plus nutrients, metal plus metal (as yet unpublished), and Cd (in progress). The detritus, natural phytoplankton and zooplankton which provide all the food supply are brought in daily by water renewal and the toxic entity in a range of sub-acute levels is introduced as daily doses to a set of tanks. These experiments are usually run for 100 days or more. By monitoring the levels of pollutant in water, suspended matter, bottom sediment, living materials and the sides of the tanks, a complete budget is secured; transfer rates into benthos and fish can be measured and the effects on the growth of phytoplankton, zooplankton, Tellina and plaice are also determined.

To illustrate the type of results obtained these workers found the 96 h LC_{50} for Cu was 750 ppb ($\mu g\ dm^{-3}$) for O-group plaice and 1000 ppb for Tellina. Copper added as Cu sulphate solution was maintained at concentrations of 0, 10, 30 and 100 ppb added Cu (above a background concentration in the sea-water supply of 3-5 ppb) for 100 days. The effect of added Cu at all levels was to progressively decrease primary production and indirectly the growth and survival of the zooplankton. Growth and survival of Tellina was reduced but the effects on plaice were more complex, if overall adverse. The lowest addition of Cu was only a few times above background and about one-fiftieth of the 96 h LC_{50}. The accumulation of Cu to various organs of Tellina and plaice was studied in detail. A very recent development along similar lines but more realistic in its lighting and temperature regime is the use of the "big bag" suspended in a sheltered limb of the sea. The experimental tanks used in the proceeding studies suffer from the drawbacks of a large surface to volume ratio, restricted and unreal light distribution and the water tends to display a larger than normal daily temperature cycle. Small-scale "big-bags" were constructed some years ago by Strickland and others (see McAllister *et al.*, 1961; Antia *et al.*, 1963) but at that time the available plastic envelope materials were far from ideal and the maximum panel size available limited their capacity. The recent development of large nylon-reinforced PVC and tough polyethylene membranes has enabled bags of 100 m^3 upwards to be made (see Davies *et al.*, 1973). Such huge envelopes more closely approximate to the real environment than any earlier enclosures. These are being used at Saanich Inlet, Vancouver, and at Thurnaig (Loch Ewe), Scotland, primarily for the study of the food chain phytoplankton to fish larvae and can be dosed to determine the various responses to specific pollutants at near background concentrations and with an almost natural food chain, all the materials being derived initially from the sea.

These developments are very expensive to set up and operate and demand a competent team of hydrographers, chemists, biologists (algologists and zoologists) and mathematicians to assemble and evaluate the great body of data that can be obtained. Their application is thus restricted to the most important and most vital pollution problems such as effects on the eggs and larvae of commercial fish species at risk. This can only be realistically appraised using a natural food chain based on phytoplankton, especially for species that spawn offshore in the upper water layers. The results of these big bag experiments are eagerly awaited. This approach is clearly a very valuable one because it begins to fill in the very wide and unpredictable no-man's land between the often artificially contrived controlled laboratory experiment and the real situation in the sea.

The zooplankton-based food chain has inherent merit and offers some advantages over the benthos-based chain. It is clearly capable of being scaled up from the tank to the big bag. The special attribute of the zooplankton based food chain is that it presents a means of simplifying the food chain to pelagic fish. For this reason, it is probably insufficient to monitor the chance introduction of phytoplankton and zooplankton in the daily water renewal as for the benthos linked model (the benthos effectively integrates any day-to-day variations) and instead the primary food source and the herbivores must be controlled either as stock laboratory cultures or as some pooled supply of plankton maintained for the purpose. The main advantages of the plankton-based model are the shorter food chain and the creation of a new sphere of study, because sandeels can be much more readily brought through their whole life span from egg to mature adult in captivity. Sandeels are exploited as the basis of a vast industrial fishery in the North Sea but are not used directly as a food resource. The more important pelagic species, herring, sprat and mackerel, have not been amenable to successful domestication in large tanks and their sensitivity to even the most careful handling for growth and feeding measurements limits their full exploitation as experimental animals.

4. *Laboratory Evaluations Including Biochemical Pathways*

The very large amount of work falling into this category would not be described by the scientists involved as any kind of toxicity assessment. For example, a great deal of work has been done on the detailed biochemistry of metals, pesticides and all kinds of other specific pollutants or pollutant groups (such as oil components). In such experiments, attention was devoted entirely to the detailed biochemical processes that handled the pollutant or were affected by it and to the associated chemical changes in the form of the pollutant. Such preoccupation was entirely right. A great deal of this biochemical knowledge in particular as it relates to the accumulation or translocation or transmutation of pollutants lies close to the core of biochemical pathways to man and to the general basis of toxicity, its function and its assessment.

This theme is simply a biological viewpoint of the biochemical arguments developed earlier and in which the specific examples of metallic inhibition metallo-thioneins the action of growth hormones pesticides and the biochemistry of hydrocarbon have been studied in some depth.

To a very large extent, biochemical mechanisms are the underlying factors that constitute toxicity as it may be displayed as death of the recipient or, as expressed biologically in many ways such as by change in growth rate, predator mortality, sexual development, reproductive success, migratory pattern, tissue and gene characteristics, resistance to disease, behaviour and ultimately by transmission to man with or without overt adverse effect.

B. Response Mainly Involving the Success of Marine Species

1. Feeding Studies, Food Chain Conversion and Growth

Despite the obvious nature of studies such as the acceptability of contaminated food to fishes and other marine organisms, virtually nothing has been published on this topic. To resolve a complaint that mussels (Mytilus edulis) contaminated with oil were useless as bait for haddock lines, samples were removed from their shells and fed to aquarium kept fish (whiting and haddock). All the fish presented with mussels (which had had oil on their shells but had remained alive) took them avidly and without ill effect. However, in tank tests in which cod were routinely fed with a measured diet of squid flesh dosed with topped crude oil, the cod, after initially accepting tainted squid for several days, most resolutely refused it thereafter. To overcome this problem the dosing with oil was done by feeding halibut oil capsules dosed with the crude oil followed by the normal diet of fresh squid.

Incidental to the tank experiments and big-bag experiments already described, it was found that some metals depressed the growth of the resident phytoplankton and even of the attached macrophytes.

Laboratory feeding experiments using Cu-enriched sea water showed a decline in the feeding rate of copepods for phytoplankton. This decline is thought to be a combination of effects of Cu on the copepods and possibly Cu-contamination of their food. Rather similar results were found in big-bag experiments using Hg.

Crude oil and petroleum products are major pollutants of the Caspian Sea reducing the biomass of the complete food chain (phytoplankton, zooplankton, benthos and fish). Sea-pike (Lucioperca) have been wiped out and in general commercial catches were reduced.

Not all the picture is black, however. In Irvine Bay in the Firth of Clyde, west of Scotland, in a region of industrial eutrophication and substantial raw sewage input, the benthos of this large sandy bay is especially rich and diverse. The area supports an unusually high population of newly metamorphosed and juvenile plaice, with probably the highest density and growth rate of any bay in Scotland. Similar beneficial effects on flatfish populations have been attributed to the southern part of North Sea, influenced by a considerable degree of entrophication. At the same time, there may be equally valid evidence for this benefication to be at the expense of other species, especially of pelagic fishes.

2. *Reproduction (Including Egg Survival, Embryonic Development, Larval Metamorphosis and at a much Latter Stage Maturation, Behavioural Aspects of Sex, Fertility and Fecundity)*

Since toxicity testing of certain aspects of reproduction must of necessity relate to periods of several days to several weeks, such tests fall between the categories of sub-acute and long-term. In as much as toxicity studies on reproduction may represent a crucial turning point in the success or survival of a species and frequently the sensitivity to pollutants is greatest at the egg, larval and early metamorphosis stages, the sequence of responses related to reproduction may be regarded as intermittent episodes in a long-term process. Between juvenile and adult no detectable response may exist in relation to an ambient concentration of pollutant that is inimical to reproductive success. Little is known about maturation, fertility and fecundity of marine aquatic species in response to pollutants (see, however, Arthur and Eaton, 1971; Wildish, 1972) but a lot is known in relation to sea birds, fish-eating birds and the higher carnivores of the sea especially for metals and pesticides. If comparison between aquatic animals and these rather specialized groups is valid, then one would anticipate some subtle adverse effects to obtain. Whether these exist is for the most part speculation at present.

Behavioural aspects of sex and fertility are to some extent related but the former has been much more studied. Perhaps the most unifying element in the many associated behavioural acts is some interference with chemoreception. The whole topic has been recently reviewed and collated by Grant and Mackie (1974) though not specifically in relation to pollution. Some aspects such as the homing of salmonids to their parent river must be familiar, the intervention of pulp-mill effluents has already been mentioned and O_2 concentration, change of temperature and H ion concentration are but a few of the many known contributory factors that alter migratory pattern, all under the stress of the reproductive urge. Some. species like herring, sprats and mackerel are noted for their migratory prowess, but many others if not all species of fish normally undertake limited migrations or aggregations related to reproduction and are eminently likely to suffer interruption of such behaviour. The species most exposed to interference by pollutants are those that have limited spawning areas near potentially adverse areas; for example, the herring in the Firth of Clyde, Firth of Forth and off industrialized regions bordering the North Sea. Other seas will have other species equally at risk; for example oyster, cutthroat trout and herring in Portage Inlet, Vancouver, B.C. (Waldichuk, 1968), a situation which has many analogies with the Firth of Forth. Species such as sprat, cod, haddock, whiting and many others spawn offshore in the upper levels of the sea remote from fixed engines of pollution but are nevertheless vulnerable to exposures; for example to oil from offshore developments or tanker traffic and the unexpected threat of an accident involving a hazardous cargo. Since many behavioural responses are manifested to concentrations in parts/billion, the ensuing sub-acute interference may be more grave than local mortalities in concentrations of parts/million or higher. Again this is somewhat speculative, but altered sex-related behavioural responses are well established for many species of fish and shellfish that are better adapted for study under experimental conditions.

The spawning products of the most important fish and shellfish of commerce are difficult to study:

(a) because they are inevitably only available from a very limited sub-population captured in prime condition at the point of or just prior to spawning,

(b) spawning is usually limited to a few weeks or months and

(c) spawning products are easily damaged during transport or storage or in aquaria.

If the success of "control" egg or larval samples is appreciably below what is normal in the field, there are experimental factors not being adequately controlled and the experiments are suspect, perhaps worthless. For these reasons the number of fruitful studies on the eggs and larvae of commercial species is rather small. Braarud (1955) investigated the hatching of cod eggs in sea waters of different origin. Hatching was impaired in water from the polluted Bonne-Fjord. Surprisingly, kraft bleach pulp waste did not affect the viability of the ova and sperm of the sockeye salmon but produced a slime that resulted in the increased mortality of the incubating eggs and alevins. Kuhnhold (1972) has studied the effects of crude oil with and without "non-toxic" dispersant and of crude oil seawater extracts on the survival to 100 h of one-day-old herring larvae. Most of this work demonstrates that dispersant-treated oil may be more toxic than sea water either with an equal amount of oil or with an equal amount of dispersant. The results for the soluble extract of crude oil do, however, demonstrate significant mortality. The extrapolation of these results to the sea is closely linked to the extent that the exact experimental conditions employed are representative. In a further study, this time of the survival and development of cod eggs (Gadus morhua) in presence of seawater extracts of various crude oils, Kuhnhold (1974) showed that that of Iranian crude was not especially lethal to the eggs but many of the larvae developed abnormally and either died before hatching or shortly thereafter. A lot more work requires to be done to see to what extent (whether orders of magnitude greater or less) such effects are already demonstrable in relation to the existing levels of "back-ground" hydrocarbon pollution and to levels prevailing as the result of a contemporary oil spill. The importance of such studies and their bearing on the success of new broods of resource species need not be stressed.

Major factors in the variation in the numbers of new recruits to the fisheries each year are various parameters of the "climate" and of the "living environment" with which they are associated. In coastal waters, this "climate" may involve the variable presence of waters originating close inshore and laden with sewage and industrial waste products. This situation has indeed been encountered in practice and has been studied by Steele and Baxter (1973), Baxter (1974) using herring-spawning products. In the Firth of Clyde, herring spawn over gravel beds on Ballantrae Bank in February/March. Some weeks later, the yolk-sac larvae are borne further into the estuary and, depending on the precise timing and course of the prevailing water movements, a variable part of the stock is brought into the plumes of polluted inshore water from Irvine Bay and from the upper Clyde estuary. From special trawls of short duration over the spawning grounds, ripe males and females are selected. The ova and sperm are collected and carefully transported to the laboratory where the eggs are fertilized and set out in shallow basins. Success rate in fertilization is normally very high (more than 90%) and after a short time the unfertilized eggs can be easily detected and removed. The fertilized eggs are then transferred to spotlessly clean sand-blasted glass plates and arranged on them in known numbers as a single layer, no egg touching its neighbour. The plates are then exposed in replicate in "control" and either a range of the same source of water dosed with pollutant or in a range of distinct water types drawn from relevant parts of the region. In this way, it was found that Cu at levels not greatly above background promoted the premature emergence of larvae which subsequently died off at a higher rate than the "controls" (Steele *et al.*, 1973). The response to the less definable variations between water types drawn from the area and outside of it also showed for waters known to be polluted a marked induction of premature larvae which subsequently displayed abnormally high mortality during the absorption of the yolk-sac. There are a number of prior experiments showing that pollutants commonly but not exclusively promote the emergence of premature larvae (see references, *ibid.*). Kuhnhold (1974) found that the stress of hydrocarbon pollution tended to delay the development of cod eggs and emergence of cod larvae. In later experiments in ayear of high estuarine exchange with the outside waters, Baxter (1974) found much less marked effects for what had been polluted areas in more normal years.

By and large, all spawning products of resource species are equally valuable as eventual contributions to the fishery, though local adult spawning stocks may represent a potential major source of income for some specialized local fisheries. For the more constrained populations of inshore benthic animals, it is not unusual for separate communities of the same species to contribute very different functions to the total population. For example, Buchannan (1966) showed for populations of the heart urchin Echinocardium cordatum that live off the Northumberland coast that only the lesser part of the community that occupied the coarse sediment at the low-water mark had any reproductive potential: those that lived in great abundance offshore did not reproduce. On a purely numerical basis, it would appear that the offshore community was the more important but elimination of the smaller and more vulnerable inshore community would eventually result in the loss of the entire population of the area. Thus the assessment of the effects of a toxic effluent must take into account such donor-acceptor dependence between "separate" communities. One must hesitate, therefore, before assuming that any part of an animal population can be dispensed with simply because it may appear to be a minor component. Incidentally, the heart urchin is a useful source of food for a number of commercial fish species.

3. *Markting Experiments and Live Boxes*

The use of tagged or otherwise marked animals held in live boxes can extend the rather formal and limited information of acute toxicity tests to encompass some or, if desired, all of the great range of "climate" and of ecological interaction. Also these techniques can be used routinely to integrate over a suitable period the variable nature of a discharge and its movement and dispersion with or without the additional stress of species competition. The technique can be varied to take into account survival at various stages of development. Shellfish have the physical robustness that suits the unavoidable minor knocks and stresses of the handling which is needed in this procedure. The hardier fish have also been tried, but experiments on fish eggs and larvae set out in live boxes with fine gauze sides were unsuccessful because of the deposition of detrital material and growth of slimy bacterial film.

In a study of the solvent emulsifier BP 1002, Perkins (1970) marked and liberated treated animals to the shore. (This very efficient oil emulsifier was used extensively during the Torrey Canyon clean-up operations. It is still used in special situations but in general the newer "non-toxic" formulations are specified. Following the accident, there was considerable interest in somewhat perplexing marine toxicological aspects of BP 1002.) The snail-like gastropod molluscs Littorina saxatilis, Littorina littorea and Tha is lapillus have been subjected to formal 24 h LC_{50} tests. L. saxatilis and L. littorea have a 24 h LC_{50} above 3000 ppm for BP 1002, and that of T. lapillus is 300 ppm. The Littorina spp. were treated with 1,8 and 100 ppm BP 1002 and Nucella with 0·1, 0·8 and 10 ppm. Despite the fact that these concentrations were about 1/3600, 1/400 and 1/30, fractions of the $LC_{50} \geqslant$ 10% mortality occurred in the liberated treated animals compared to handled but untreated controls. Furthermore, the growth of the Littorina spp. was significantly reduced and even after 22 weeks that of L. saxatilis treated at a 1 ppm BP 1002 had not regained the normal rate. Treated Litttorina saxatilis was noted to be preferentially consumed by the shore crab, Carcinus meanas. Gribbon (1973) worked with the oyster Ostrea which he treated with a range of diluted oil emulsifier formulations. The treated oysters were first retained in live boxes and then laid on an appropriate ground and left unmolested for almost four years (3 years 42 weeks) after which they were lifted and examined. Mortality significantly greater than "control" occurred in those exposed briefly to BP 1002 at only 1/128 fraction of the 24 h LC_{50} and growth was inhibited for test concentrations above one-eighth of the 24 h LC_{50}. The condition factor, however, was greater in the animals treated with BP 1002 than in the controls but for the "non-toxic" BP 1100 no detectable survival or condition effects were found. For all the test oysters the condition factor was above the standard acceptable for marketing.

Quite a number of experiments have been carried out on the recovery in the laboratory of small invertebrates that had been exposed to near lethal doeses of oil emulsifiers, in the laboratory, in simulated field trials and from real clean-up situations (Perkins, 1968; Crapp, 1970; Griffith, 1970; Portmann, 1970). With fish, however, such tests may be carried out more meaningfully in an open situation in large live boxes in which the fish can be exposed to oil and oil dispersant applied in a manner simulating that of a real oil spill clean-up operation (Tracy *et al.*, 1969).

One of the many and obvious restrictions of laboratory simulations is the difficulty of providing the test organism with the normal range of food in normal amounts, which may not be serious for smaller invertebrates but makes the study of the larger ones like oysters and whelks rather unrealistic. For some species, recovery subsequent to exposure to a near acute dose of pollutant may be studied over a truly representative part of the animal's lifetime. Live boxes have been used to study mortality, growth and condition of oysters during three years subsequent to treatment with oil emulsifiers (Perkins and Gribbon, 1972; Gribbon 1973) and Perkins (1970) also studied treated winkles, Littorina, and dogwhelks, Thais, liberated to the shore. Although liberation is clearly the ideal technique, even these rather sluggish animals roam about or are carried away from the site of release and with time their recapture becomes progressively more difficult. Although one gains confidence and credence in such results derived from the recovered labelled animals, the recapture of only a small or moderate proportion of all the control and treated animals provides only incomplete and perhaps a biased part of the total information and thus introduces an element of uncertainty into the interpretation of the results. If the species chosen is non-mobile or if slightly mobile unlikely to become impossibly hidden or widespread, this test may be advantageous and trials may be as long as 26 weeks (Perkins, 1970). Statisticians would recommend increasing the number of marked animals and perhaps replication of sites; the latter may be helpful, but there is a strict limit to the number of a species that may responsibly be interfered with at any one site and frequently the numbers required are already as high as practical if a range of treatments is to be covered.

Line boxes, marks and tags have been quite extensively used to study effects on freshwater fish. So far as I know, no similar experiments have been deliberately performed on marine fish. Many tagging experiments have been carried out on plaice, lemon sole, haddock, whiting, cod, halibut, and other commercial species of the North Sea and elsewhere. Invariably these experiments have been evaluated in relation to migratory routes, growth rate, survival, maturation, population dynamics, checking ring formation on scales or otoliths and other aspects of fisheries biology (Jakobsson, 1970). Comparable studies have been made on crabs, lobsters and other important shellfish for similar reasons. Also transplantation experiments have been made to explore growth rates of a unit stock in various habitats. One wonders to what extent some of these experiments were affected by pllution. By comparing detailed analyses of the gains of toxic metals or pesticides in recaptured tagged fish or shellfish, it may be possible to get some idea of the impingement of such toxicants on the survival, growth and reproduction success of marine species at current background exposure concentrations in water and food. There are some foreseeable snags in such an approach and the resources involved are pretty demanding. The operation is, however, completely harmonious with normal tagging or transplantation work and could with benefit be incorporated with future programmes of this kind to the mutual benefit of fisheries biologist and toxicologist.

C. Field Surveys and Other Studies Designed to Evaluate Ecological Interactions

The studies of biological response in relation to man to a major extent are looking for some effect perhaps many years after they have begun to happen or an explanation or quantification of the mechanisms involved; the studies on the success of single or a few marine species are more immediate and provide fairly simple means of assessing long-term response in relation to a number of possible criteria. Neither of these approaches tackles the practical problem of evaluating what an existing pollutant is doing to the entire surrounding environment or is helpful and reliable in anticipating the probable general effects of a proposed new discharge. The field survey is the most direct but the least academic method of appraising long term ecological response and it can be very demanding in effort devoted to practical field work and to interpretation.

The simplest way to find out if a pollutant or a discharge is harmful to a receiving area is to introduce some to a small test area and see what happens. It is at once obvious that some tests such as dumping cyanide waste into a rock pool or applying oil to a patch of salt marsh are capable of yielding possibly meaningful results but, for liquid discharges to inshore waters, it is much more difficult to learn anything useful in this way. Thus only certain problems can be studied by direct application of the pollutant. There are however, two ways of tackling long-term appraisal of the effects of liquid discharges to the sea, (a) field observations of water, sediment and marine life in relation to comparable existing discharges or (b) the modelling of liquid discharges, existing or proposed, using pilot-scale tanks or creeks and determining the response of simplified or actual ecological groups.

This is the basis used here for the classification of long-term studies relating to field trials.

1. On Site Application of Pollutants

The application of a pollutant onshore, inshore or offshore can be deliberate or it may be accidental. It is easier to exploit the deliberate application but on occasion the accidental introduction can be exploited as a field trial, *e.g.*, much was learned about the effects of major oil spills from the Torrey Canyon and other ships (Smith, 1968; Wafra, Day *et al.*, 1971) and from the loss of control at the Santa Barbara oil well (Straughan, 1971). It is difficult to be suitably prepared to exploit accident situations involving hazardous cargoes in general because the location may be remote from laboratories with appropriate facilities, the pollutant may not be known very well and ready means of quantification may be lacking or laborious; for oil spills, however, several countries have prepared outline official programmes of research to be fitted in without interference to any clean-up operations. It may also happen that accidental release occurs in a region that has recently been surveyed prior to the incident and, in consequence, the ensuing effects can be assessed in the short-term or for years after as may be desirable. For such accident appraisals, it is necessary (a) to secure samples of the pollutant for detailed analysis and for any controlled tests that may be enlightening; (b) to obtain reliable estimates of the amount released and to secure water samples to determine its distribution. Much will depend on the priority of current programmes as against an *ad hoc* investigation and the spare capacity available to tackle adequate surveys and studies such as those using marked animals and live boxes suggested above.

An outline has already been given of the work done on salt marshes to determine response to single and repeated oilings and on elucidating a wide range of ecological repercussions. These are excellent examples of essentially simple practical approaches to determining acute, sub-acute and long-term effects of oil spillage on the coastal environment, of which the salt marsh is particularly appropriate for study in this manner and incidentally is also particularly susceptible to long-term damage.

Boaden and Bleakley (1973) studied the effect of surfactant and surfactant with solvent applied in normal dosages to the intertidal zone of a sandy shore and its typical population of meiobenthos. They found that the maximum mortality at a given level of dosage depended on the time interval between the moment of application and the first lapping of the returning tide. Areas heavily dosed with surfactant were gradually recolonized during 28 days subsequent to treatment. Recovery in a sandy environment depends initially on flushing out the pollutant and recolonization will then progress depending on the stability and grade structure of the sediment. It would be of value to build up a body of knowledge of this type for a wide range of pollutants and a wide range of shore deposits and thus provide some basic indications of the residence time and relative toxicity of pollutants for a range of typical beach situations to go along with the more ephemeral but much better studied effects of pollutants introduced into the water.

In the routine evaluation of pesticides and agricultural chemicals, generally it has been a world-wide practice to conduct plot trials for the purpose of evaluating their short-term effects and persistence (equivalent to long-term effects). Sometimes steps are taken to extend these trials to lakes. ponds and artificial or diverted streams to see what happens to freshwater fish, food organisms, predators and plant life. Seldom is the past has any attempt been made to foresee the consequenencs of the release of the toxic product (or its residue) to the marine environment. This is surely a deficiency that must be rectified in future, by uniform legal enforcement if need be, for all synthetic and novel products that may come to be used on a large scale whether for agricultural purposes or any other.

2. *Field Observations in Relation to Existing Discharges*

The limitations are pitfalls of the evaluation of field observations in relation to existing pollution are so fundamental that they are worth repeating though they may appear obvious and are familiar to many. At face value, what is aimed at is to use the knowledge gained from surveys of one area known to be exposed to a certain pollutant to foretell the effect of another similar discharge elsewhere. This is perhaps rather like applying, say, British precepts in child care to the upbringing of a child in Canada or might it be in Central Africa or the Arctic Circle. Admittedly, this illustrative example is more complex but the contrasts in environment are real enough. To some extent also, it is like using Victorian concepts of child care a century later in the same family, generations later in the same city because the environment also changes with time as well as in space.

There is no way round the differences in marine life between different areas and countries and all that can be done is to test significant selected species against as relevant a pollutant sample as can be obtained and thus derive a set of relative toxicities under suitable conditions of temperature and salinity. With regard to temporal variation it is often quite nonchalantly recommended that a base-line be established preferably and/or a reference area of "control" which will resemble the reception area but will escape the effluent. However, as J.R. Lewis has repeatedly pointedout base-lines may be particularly difficult to establish because of natural fluctuations and the long-term fluctuations characteristic of the species or of the community that surrounds it may greatly exceed that which may be apparent in the time made available for pre-or post-commissioning surveys. This has been exemplified frequently for the inshore environment but it is evident also that the apparently more physically and chemically uniform offshore environment is at least as variable in some respects. Regarding the starfish Asterias forbesi observed by Loosan off (1961) from records covering the period 1934-1957, there were years of markedly increased abundance (1934, 1939, 1948, 1949, 1950 and 1957) and the average number and size/station was highly variable (Fig. 3.3).

Similar widely varying abundance has been found for the oyster Ostrea edulis (Waugh, 1966), the bivalve mollusc Tellina tenuis (McIntyre, 1970; Perkins 1972d, e) and Macoma balthica (Segerstrale, 1962; Perkins, 1972d, e) Marked natural changes in the zooplankton off Plymouth between 1930 and 1970 are recorded by Russell *et al.* (1971) and variation with an imposed major trend in North Sea plankton by Colebrook *et al.* (1972).

Within the time-scale of these variations, there are of course seasonal and other short-term variations; many of these are integrated and largely disappear when longer-lived residential species are studied, though at critical periods, such as during early larval stages, the presence or absence of planktonic components may exert a major effect.

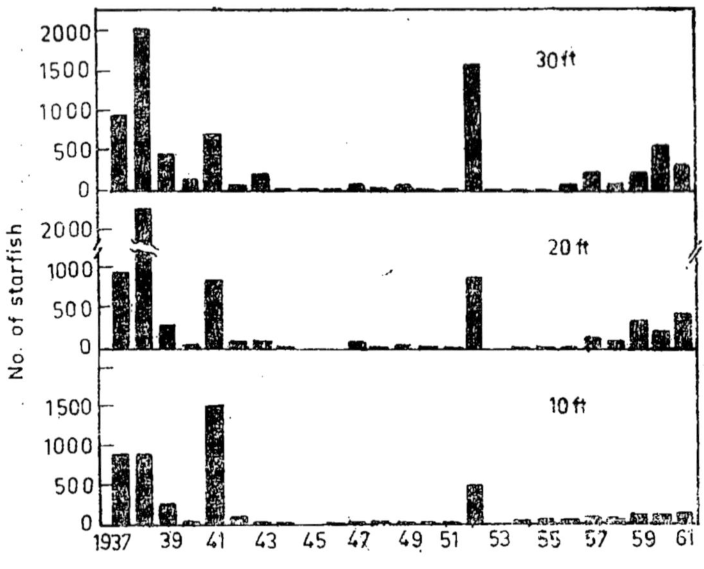

Fig 3.3: Average number of starfish set per station recorded in different years on 100 oyster shell-surfaces at stations located at 10, 20 and 30 foot depths, Long Island Sound, 1937-1961 (after Loosanoff, 1964)

Examples of long-term environmental studies in relation to pollutants are numerous and some with regard to pulp mill effluents and acid mine wastes are very well documented for fresh water. A great many relate to sewage outfalls of which many are parochial and depend almost exclusively on the very local conditions around outfalls, on the nature of the outfall, on the population served and on associated amenities. Others such as the collective discharges to the bay at Marseille (Bellan, 1970) demonstrate the damage suffered by marine life over a wide area when the discharge is voluminous and dispersion conditions are less than normally favourable. Similarly, the changed environment associated with the dumping of sewage sludge over a long period of years has been described by Pearce (1970). In general, all situations dealing with the long-term introduction of wastes rich in particulate matter are complex Perkins 1972d, e).

Sometimes, the scientist conducting field studies related to one pollutant is confronted with the uncertain influences of past events (unnatural ones) on the present situation. This can happen in relation to the earlier dumping of solid and certain metals inert wastes. Solid wastes may often appear to be insoluble when exposed to water in a beaker; this may not be true for the same material in contact with virtually unlimited sea water. For example, a chemistry student learns that Ca sulphate is insoluble and may use it as a gravimetric assay under defined conditions, yet 2 g will dissolve in 1 litre of distilled water and 4 g in 1 litre of sea water (Shaffer, 1967) which in the long term is quite enough to safely mobilize (partly dissolve and disperse) vast amounts discharged as gypsum from fertilizer manufacture. Other materials such as shale wastes and coal slag may give rise to toxic metals and organic compounds at least initially on discharge to the sea.

Such gradual slow release of toxic contaminants does not reveal the full extent of the problem.

In the very early stages of the spread of industrialization near Workington, Cumberland, an industrial town on the north-west coast of England, hot slag was poured directly on to the shore from the blast furnaces where it solidified to form beds of an artificial rock called "siagcrete" (Perkins, 1972d). These discharges to the shore were made down to and below the mean tide level. A century later, the original "stacks" of slagcrete show very clearly an advanced degree of erosion. It has been shown by chemical analyses of slagcrete and of the intertidal sediments all along the shore northwards to Silloth that a large part of these sediments has been derived from eroded slagcrete. Detailed analysis of this material reveals significant concentrations of heavy toxic, metals and, from a series of biological surveys during the past 14 years, it is apparent that many elements of the intertidal infauna become progressively impoverished as Workington is approached. One essential difference found was that at extreme low tide level where the almost constant presence of water removes any leachate, despite the fact that conditions are turbulent for much of the time, the fauna there is abundant and healthy. Offshore, admixture of the pulverized residues of slagcrete with the substratum creates a range of more mobile sediments and this instability is primarily responsible for the impoverishment of this terrain. This example illustrates how long-lived and long-forgotten materials dumped on the seashore may result in irremedial change not only over one lifetime or one century but for the foreseeable future.

These findings may probably be valid for any dumpings of any foreign inert material over any natural substrate and includes physical influences on local water movement as well as any physical or toxic effects on the native fauna.

Other hazards may arise from the dumping of inert materials in areas of active sediment migration. Typical situations have been studied in relation to the Thames estuary, Bristol Channel-Severn estuary, Solway Firth, Gota-Alf (Sweden), Yaquina Bay, (Oregon), Cooper River (South Carolina) and many of the world's major rivers (Perkins, 1974). One important finding is that sediment dumped within the active area (which may extend miles from the parent tidal estuary) can result in the prompt return of the material to the shore or saltmarch and the morphology of the channels and shoreline may be changed. This may affect for example characteristic saltmarsh zones.

The physical mechanisms of coastal photography sediment balance and transport are complex and a detailed discussion is outside the compass of this chapter. The importance of these mechanisms is that they can provide a means of physical reconcentration of certain fractions of dumped material. Asymmetric distributions with possible concentration arise from the winnowing and re-settlement of particulate materials of favourable size, density and consistency under the action of tidal or tidal plus river streams. Mobile materials are moved in characteristic patterns by topography and asymmetric tidal forces and can infringe on established flood and ebb channel developments which may become altered depending on factors such as the degree of competition of foreign with existing materials. In turn the channels, when disturbed, affect features such as the exposure of the shoreline to the incoming tide.

The onshore transfer of the relatively buoyant components is the most familiar demonstration of such effects at one end of the range. It is commonly found that plant remains, sewage entities, small litter seaweed and drift objects such as sea-bed drifters and radioactively labelled deposits all tend to drift in an asymetric manner to be accumulated in characteristic patterns. These situations are also favoured by juvenile polychaetes such as Arenicola and bivalve molluses such as Macoma (Perkins, 1974). Thus both larva and juveniles and actual or potential food materials or favourable substrates tend to be brought together by a mechanism that is essentially beneficial in one evolutionary sense yet in a polluting situation may place the most vulnerable stages of the species at increased risk (Connor, 1972).

Similar considerations should be given to the very different problem presented by very deep-sea dumping which is almost exclusively of the worst types of polluting wastes. At first sight the deep seas are vast and unexploited and, therefore, some would argue attractive alternatives for the disposal of dangerous wastes which are "impossible" to recover, or does this mean "uneconomic" to recover. There are several relevant comments to be made. Jannasch (1970) states that the high degree of stability in the physical environment indicates very slow water movement, that the specialised deep-sea life is unadapted to environmental change and that deep-sea populations including bacterial populations are low and any biodegradation will be correspondingly slow. As Allen (1973) points out, many of these deep-sea animals have a prolonged life history which renders them particularly prone to pollutants and their replacement rate is very small. At intermediate depths, particularly in waters just beyond the continental slope, lie the only known unexploited stocks of fish potentially useful to man. The have been described as man's last food reserve. It is important that these are not polluted before it becomes necessary to exploit them. (This is rather a remote diversion from the main topic of long-term biological assessment, but its urgency will rise steeply as the problem of finding suitable deep well disposal sites on land becomes more intense.)

Field studies dedicated to nominated sites suffer from yet another shortcoming, that many species have stages of their life cycles widely separated in the sea. Among inshore benthic species whose settlement of early forms may not take place at the site finally occupied by the adult are the mussel Mytilus (Bayne, 1964), many species of oyster, the cockle Cardium (Waugh, 1960) and many others. Arenicola (Newell, 1948) also has differing distributions of adult and juvenile stages. These separations may serve to reduce competition between adult and juvenile. Offshore dwellers with inshore phases are the salmonoids, many of the races of the North Sea herring, some skates and rays and many other fish species that, to a lesser degree, spend their early life towards the land. It is very important to include both groups, especially the latter which may more easily be overlooked, when enumerating the species at risk in a discharge area.

Little seems to have been said favourably about field studies or surveys. Biological surveys have undoubted value for determining the effects of a specific pollution situation in the context of the whole gamut of pollutants that may be involved (as for example in New York Bight) and of the important species concerned. In this circumstance, the biological survey is akin to the more formal Biotic Index used in river assessment. To be of more general interest it is better to have only one main source pollutant to deal with such as the various estuarine surveys related to the Canadian or the Swedish pulp-mill industries. Commonly a lot of effort is wasted in field surveys because it is not realized or it is not apparent which parameters are the ones that really matter. For example, for pulp-mill effluents, experience proves that chemical surveys (temperature, salinity, O_2, BOD and nutrients) tend to follow a well-marked set of annual cycles that are repeated every year with some minor variation due to (a) effluent loading and (b) climatic factors especially rainfall. Only in systems where the effluent plays an unduly large part, and for this reason should not have been accepted in the first place, are such surveys necessary several times a year every year. Similarly, it is seldom worth while putting out a great deal of continued effort on plankton study if the area is reasonably healthy. For pulp-mill effluents, losses of fibre and perhaps paper fillers, etc, are frequently cumulative and studies on the rate of sedimentation and distribution of sediment are at least one stage nearer to the key parameters affecting the biological system than any measurements in the soluble phase.

At mills sited on coastal plain estuaries and on open coasts where dispersal is vigorous, other factors may predominate. At such sites, the key interest may be in the rate of cellulolytic attack on suspended fibres.

Reports are awaited of exciting new studies on the roles of certain bacteria and protozoa growing on and near pulp fibres in bottom deposits. The rate of breakdown has been shown to depend on factors such as temperature, oxygenation, water exchange and nutrient supply. Specific bacteria and protozoans are believed by the scientists at the Scottish Marine Biological Association, Dunstaffnage, Oban, to be the primary agents, singly or working together, involved in the hydrolysis of the purified fibres.

Associated with these studies are parallel studies on the growth, survival and species composition of the related benthic populations in the field and in laboratory tanks.

On the biological side, it is often only necessary to evaluate the main groups which make up the benthic populations of polluted and relatively unpolluted parts of the system; further detailed study of all the species involved may have some academic value but contributes nothing to the understanding of the overall response. The measurement of redox potential in bottom cores may often show some parallels with benthic faunal composition. Similarly, with regard to hydrographic studies, one or two well-designed dye or labelled fibre studies are perhaps initially more expensive and more demanding than the usual thermometer and bottle work but the results have infinitely more direct bearing on the aspects of movement and dispersion that matter.

To sum up, the academic approach to field studies involving detailed hydrography, current profiles, models, chemistry of pollutant and the usual O_2 and nutrient parameters, phytoplankton, zooplankton, benthos and pollutant toxicity, may be of no benefit unless the major problems involved have been identified and the means of solving them have been sought. Pollution evaluation is either a practical subject or it is nothing. In the words of Cole (1970)

> a different approach may be preferable by which the effects of potentially toxic pollutants on the processes of growth, physiology, behaviour and breeding are established in controlled laboratory experiments and are then looked for in the field. Where gradients in the concentration of pollutants in the sea can be found, as, for example, off the mouths of rivers carrying sizeable quantities of easily recognizable materials such as metallic wastes, it should be possible to establish in sedentary animals and plants a comparable gradation in pollution effects, for example in growth rates, behaviour, maturation of gonads, etc. From such a confirmation of precise laboratory experiments and field studies of polluted situations, it should be possible to build up an appreciation of the ecological consequences of pollution and so reach sound judgements regarding the capacity of the marine environment to accept waste without damage to living resources or productivity.

This approach is in a sense a counsel of perfection since it presupposes that the hard-core information needed to make these judgements can readily be acquired and the ensuing interpretation will be unequivocal and capable of practical application. Undoubtedly, field studies will go on in future but the technique must be used incisively, speeded up as may be reasonable, and the prime issues whether they be human health, fisheries (directly or via food chains) or amenity must decide the optimum approach including relevant means of assessing toxicity and biological response. The time-consuming, labour-intensive ecological field survey can be so slow and so non-definitive that industrialists and administrators might be apprehensive about commitment to such expensive exercises. Finally scientists, administrators and industrialists can all make mistakes however good their aims and intentions. Conditions and consents for discharges tend to be afforded a "once for always" certificate of approval. In practice any or all of these three participants are aware of aspects of the consent terms that could be altered with benefit to all, but how seldom does this materialize. From the point of view of scientific error or ignorance it must always be possible to require and obtain immediate and positive action when all does not appear to be going well.

3. *Pilot Scale Models of Simple Ecological Groups*

Because of the limitations of the field surveys just discussed and because of a need to assess long-term biological response on communities rather than single species and to provide a means of testing new polluting discharges, recourse has been made to pilot scale models of real situations including representative ecological groups.

These models are of basically two kinds, static models or rather models maintained at a static level of contamination, and dynamic models in which the effluent and the receiving waters vary more or less in a natural manner. A number of the systems discussed under Section A3, above, are representative of the first type but have rather limited ecological coverage. The use of labelled animals or live cages is akin to a dynamic model but there is no measure of the exposure to pollutant.

Most of the relevant work has been done in North America, notably in relation to pulp-mill effluents, but toxic metals, pesticides and other continuing or presistent pollutants have also been investigated, almost exclusively in fresh water. The models have ranged from artificial ponds, tank and flow-systems to controlled streams into which known doses or known rates of pollutant have been introduced (see for example Warren and Davis, 1967; Gannon and Phaup, 1967; Davis and Warren, 1968; Burrers and Luhning, 1969; Gilderhaus *et al.*, 1969; Whitworth and Lane, 1969).

The technique is fairly costly, very demanding in relation to maintenance and running, and can be labour-intensive unless suitable instrumentation for the requisite parameters being studied is no the market. It is an approach favoured by mathematical modellers and complements (if it seldom accompanies) the physical modelling of estuaries, harbours and other restricted waters for water and sediment movement and exchange Both types of model are subject to rather similar limitations of interpretation.

III. BIOLOGICAL RESPONSE AND WATER QUALITY EVALUATION

To recapitulate, assessments of the toxicity and biological effects of effluents and identifiable substances released to the marine environment have among their objectives the prevention of death to human and marine life, assurance of the continuation of the complex food web and of the general health and productivity of marine life from which man derives great benefit. It must be clearly understood that waste disposal is not necessarily harmful though harm may arise on occasion and over a limited area. One of the objectives of toxicity testing is to ensure that waste disposal is on a scale that can be absorbed, and longterm biological assessments form an essential part of this testing procedure.

The appraisal of the capacity of the system of receiving waters to accept wastes to a predetermined acceptable level of contamination is seldom explored. In part, this is inhibited by the realization that discharge controlled in such a manner as to create nearly gradient-free distribution of pollutant is either impossible or would be monstrously expensive except for very special disposal sites. In practice, therefore, some clause excluding a conventional immediate dilution area must be foreseen in relation to the volume and strength of discharge, assuming that the discharge meets conventional discharge limits; factors such as 100 or 500 effluent volumes would seem reasonable.

Most assessments for consent to discharge relate to specific discharges in isolation but, in relation to the concept of water quality standards, it becomes necessary to involve all the sources of pollution entering the receiving water. If an adequate evaluation of water quality beyond the immediate discharge zones can be established, it becomes feasible to define an at least minimal site capacity which would preserve an acceptable quality for the biota. To the chemist, there is considerable attraction in the summation of biological water quality as a Biotic Index; to the biologist the use of water quality assessment represents a summation of chemical and physical parameters. There have been many attempts to establish criteria for water quality evaluation in fresh water (Cottam, 1959; Cleary, 1960; Smith, 1961; Tarzwell, 1961, 1962a, b, 1963; Surber, 1962; Horton, 1967; State of Washington, 1967; F.W.P.C.A,, 1968; and others). Water quality standards for the sea have been attempted by State of Washington (1967), F.W.P.C.A. (1968), Tarz-well (1969), Batelle (1971) and perhaps a few others. It is self-evident that standards have to take into account physical (*e.g.* heat) and all forms of chemical pollution. To set standards that apply only to the commonest effluents from land-based discharges has many weaknesses (Cairns, 1968). Another limitation is the formula used in summing up the effects of many different parameters. For example, consider the problem of evaluating the total human hazard arising from radioactivity to a typical resident in a town somewhere. Radiation in some slight measure reaches the subject from the air he breathes, by contact with clothes and skin, from drinking water and beverages, from landbased food, from sea food, from the sky and from the material of his house, etc. The total effect from earch source can be determined and these values can be set against limiting standards set by I.C.R.P. in 1955 for each source and also for the total does received. Unlike radioactivity, there is no simple single measurement that can generalize the broad spectrum of physical and chemical marine pollutants. Instead of one single measuring unit, the evaluation has to be made on satisfactory compliance with many standards, and these standards must be valid for the biota of the receiving area, not just one typical resident.

This critical issue raises once more the measures needed to ensure that conditions of water quality described by Cole (1970, above) are fulfilled. Find by survey the more important components of the overall population that are or are about to be exposed to polluting influences. Determine the most relevant acute and long-term biological responses. Determine the probable pollutant distribution and variability using existing discharges or a marker substance. It then becomes possible to define a concentration in the major part of the discharge area that is truly safe for the health and productivity of the major sector of the biota. Since the safe concentration has been determined by a positively oriented toxicity survey, the maximum permissible burden is related to a known magnitude of input, known dilution and transport. If particulate matter is present, separate account will have been taken of its proclivity to reconcentration.

To take into account operational malfunction, it is possible from statistical data for each industry to foresee a probable rate for spills and discharges of effluent outside normal range of composition. All this is possible but somewhat idealistic for the normal run of industry, though such statistical appraisals are routine for nuclear plant. A suitable margin is often calculable, especially for upsets during the initial setting-up of factories, and a safety factor is usually applied to the normal range of toxic parameters in the effluent.

Once a new major effluent has been initiated, it should be the subject of continuing surveillance to assess the validity of the judgement made in the precommissioning period.

Certain consent authorities are committed to regulate discharges solely on the analysis of toxic components and will pass or fail an effluent on concentration limits. This is palpably unfair and unrealistic since it does not take into account the magnitude of the discharge or the capacity of the receiving water to accept it. Equally, these may not be appropriate concentrations for the more sensitive elements of the biota and some modification of the effluent or its mode of discharge may be necessary. Clearly there is room for flexibility in approach in marine pollution management. Water quality criteria offer a possibly more valid solution to such problems.

A plea has been made by U.N. groups for a movement towards achieving a unified set of water quality standards for application world wide. No single set of standards would of course meet the very different environments of polar, temperate and tropical waters and any set of working standards chosen as a starting point would have to be modified to establish criteria at least with zoning for latitude and with perhaps special standards for inland waters and enclosed or semienclosed seas.

The water quality standards first defined by F.W.P.C.A. in Water Quality 1968 and outlined by Batelle (1971) are American for American situations and will be used here for debate and development. In the following paragraphs these standards are outlined as they might be applied more generally. Without any intended disrespect to the originators their wording has been anglicized and certain somewhat turgid modes of expression have been removed in order to preserve a single English style in this book. The units employed have been metricated in line with scientific practice. With due apology also comment has been made on the scope of application of such standards universally, being aware that these standards were not originally destined for such treatment and were, as here, meant as a basis for discussion.

A. Criteria for Ordinary Sea-Water Parameters

1. Salinity

For the protection of estuarine organisms, no changes in channels, in the basin geometry of the area, or in freshwater inflow should be made that would cause permanent changes in isohaline patterns of more than $\pm$ 10 per cent of the natural variation.

Comment. No basis has been set for the time scale of salinity variation. For some estuaries, if you wait long enough, the water might be either fully fresh or fully salt and naturally enough the resident population must be capable of tolerating such seasons. This stricture is therefore more applicable to sites where the normal salinity range is naturally small.

2. Water Circulation

In view of the requirements of estuarine organisms and the

nature of marine waters, no change in basin geometry or fresh-water inflow should be made to tidal tributaries to alter current patterns in such a way as to cause adverse effects.

Comment. As this recommendation stands it could outlaw all land fills and dredging operations. There should be some volume limit related to the dimensions of the area within which the operation is planned to take place. In many circumstances, the economics of land use benefit or of shipping requirements will far outweigh any local fishery on amenity involvement. If the relative dimensional change is small and conservative this should delineate fishery or amenity loss to a first approximation.

3. Temperature

In view of the requirements for the well-being and production of marine organisms, it is maintained that the discharge of any heated waste into any coastal or estuarine waters should be closely managed. Monthly means of the maximum daily temperatures recorded at the site in question and before the addition of heat of artificial origin should not be raised by more than 2°C during autumn, winter and spring, nor by more than 1°C during the summer. The imposed rate of temperature change should not exceed 0.5°C/h above ambient.

Comment. The original goes on to define "summer", etc., for various American localities. If these restraints are omitted, the clause becomes more applicable to the general situation world-wide, though there may be cause to review the precise temperature burden valid for different latitudes.

4. Hydrogen in Concentration (pH)

Materials that extend normal ranges of pH at any location by more than ± 0.1 pH unit should not be introduced into salt-water portions of tidal tributaries or coastal waters. At no time should the introduction of foreign materials cause the pH in a saltwater regime to be less than 6.7 or greater than 8.5.

5. Dissolved Oxygen (D.O.)

For the protection of marine resources, it is essential that O_2 concentrations be sufficient for survival, growth, vigour reproduction and the production of a suitable crop. To attain this objective, it is recommended that dissolved O_2 concentrations in coastal waters, estuaries and tidal tributaries be as follows.

1. Surface dissolved O_2 concentrations in coastal waters should not be less than 5.0 mg dm^{-3} (ppm), except when natural phenomena cause this value to be depressed.
2. Dissolved O_2 concentrations in estuaries and tidal tributaries shall not be less than 4.0 mg dm^{-3} at any time or place except in dystrophic waters or where natural phenomena cause this value to be depressed.

6. Crude Oil and Petroleum Products

No oil or petroleum products should be deliberately discharged into estuarine or coastal waters in quantities that

(a) can be detected as a visible film or sheen or by odour,

(b) cause tainting of fish and/or edible invertebrates,

(c) form an oil-sludge deposit on shores or bottoms of the receiving body of water or

(d) become effective toxicants according to the criteria recommended in the "Toxicity" section.

7. Turbidity and Colour

No effluent which may cause changes in turbidity or colour should be added to, or discharged into, inshore or coastal waters unless it has been shown that it will not be deleterious to aquatic biota.

8. Settleable and Floating Substances

Since it is known that even minor deposits may reduce productivity and alter the benthic environment, it is recommended that no materials containing settleable solids or substances that may precipitate out in quantities that adversely affect the biota should be introduced into estuarine or coastal waters. It is especially urgent that areas serving as a habitat or nursery grounds for commercially important species lobsters, crabs, oysters, clams, scallops, shrimps, prawns, flounders, salmonids and other pertinent species and also the eggs and larvae of fish and bottom forms be protected from any infringement of natural conditions.

Comment. This clause has a rider that permissible limits of settleable solids and floating materials cannot be expressed quantitatively at present. In view of this, the clause goes on to prohibit "any infringement." Clearly this cannot be realistically enforced and is somewhat confounded when it is recalled that rivers introduce millions of tons of settleable and floating solids annually.

A possible solution would be to base a standard for settleable solids on

(a) the increased annual burden of oxidizable carbonaceous material/m^2 and

(b) an upper limit for inert material introduced/m^2 p.a., both relative to the existing natural burden (or if the area is already polluted, above some pre-set value).

Floating materials not derived from natural inputs would be better controlled by an embargo on dumping.

9. *Tainting Substances*

To prevent tainting of fish and other sea foods, substances that produce taste and off-flavours should not be present in concentrations above those shown to be acceptable by means of bioassays and taste panels. Experience has shown that test organisms should be exposed to the materials for two weeks at selected concentrations to determine the maximum concentration that does not produce noticeable off-flavours as determined by organoleptic tests. Cooking should be done by backing the material wrapped in aluminium foil.

10. Plant Nutrients and Nuisance Organisms

(Including the events leading to eutrophication, excessive growths of attached and other algae including the producers of neurotoxins)

The ecological factors most often associated with nuisance growths result from natural changes in the temperature and salinity regime and thereby increased nutrients. Accessory factors may directly or indirectly affect the response of the organisms to these factors. Increase or decrease in water-column stability and, indirectly, its effect on nutrients in the photic zone have also been found to be important.

To maintain a balance among nutrients and a balanced biota most conducive to desirable primary production, it is recommended that:

1. No changes should be made in the basin geometry that would result in adverse effects on aquatic life. For example, studies should be made before dams are erected, water diversion projects are constructed or dredge and fill operations carried out.

2. The extent of artificial enrichment of the marine environment from all sources should not cause major quantitative or qualitative alteration in the flora. Production of persistent blooms of phytoplankton, whether toxic or not, dense growths of attached algae and higher aquatic plants or any other sort of nuisance that can be directly attributed to nutrient excess or imbalance should be avoided. Because these nutrients often are derived from land drainage, special attention should be given to the avoidance of the use of excessive fertilizers and to pursue good soil management.

3. The naturally occurring atomic ratio of NO_3-N to PO_4-P in a body of water should be maintained. Similarly, the ratio of inorganic P (orthophosphate) to total P (the sum of inorganic P, dissolved organic P and particulate P) should be maintained as it occurs naturall. Imbalances have been shown to bring about

a change in the natural diversity of the desirable organisms and to reduce productivity.

Comment. In line with recommendations there would be some justification in setting the allowable limit of major nutrients at about 3-5 times the winter background concentration in temperate waters; for regions where winter values are very low because of continuous high plant production, the limit could be set at those concentrations not giving rise to accumulation or distorting the normal flora. A lower factor may be desirable for cold regions.

B. Recommendations Regarding Toxic Substances

1. General Recommendation

In the absence of toxicity data other than the 96 h TL*m*, an arbitrary application factor of 0.01 shell be used as the criterion of permissible levels.

Additional chronic exposure tests will be conducted within a reasonable period to demonstrate that the estimated maximum safe levels as indicated by the 96 h TL_m and the application factor do not, in fact, cause decreases in productivity of the test species during its life history.

Comment. The relevance of such biological appraisals is dealt with on pp. 521-535.

2. Biocides

The biocides are grouped according to their relative toxicity to the shrimp, one of the most sensitive species of marine crustaceans. Criteria are based on best estimates in the light of present knowledge and it is expected that acceptable levels may be changed as the result of future research.

(*a*) *Group A*. The following chemical pesticides are acutely toxic at concentrations of 5 μg dm^{-3} and less. On the assumption that 0.01 represents a reasonable application factor, it is recommended that environmental levels of these substances not be permitted to rise above 50 ng dm^{-3}. This criterion is so low that these pesticides could not be applied directly in or near the marine habitat. The 48 h TL_m is listed for each chemical in parts/billion (μg dm^{-3}).

Organochlorine pesticides

Aldrin	0.04	DDT	9.6
BHC	2.0	Dieldrin	9.3
Chlordane	2.0	Endosulfan	9.2
Heptachlor	0.2	Methoxychlor	4.0
Endrin	0.2	Perthane	3.0
Lindane	0.2	TDE	3.0
		Toxaphene	3.0

Organophosphorus pesticides

Coumaphos	2.0	Naled	3.0
Dursban	3.0	Parathion	1.0
Fenthion	0.3	Rounel	5.0

(*b*) *Group B*. The following types of biocides are generally not acutely toxic at concentrations of 1 mg dm^{-3} or less. It is recommended that an application factor of 0.01 be used and, in the absence of acute toxicity data, that environmental levels of not more than 10 μg dm^{-3} be permitted.

Other biocides including some pesticides

Arsenicals	2, 4, 5-T compounds
Herbicides	Phthalic acid compounds
Carbamates	Triazine compounds
2, 4-D compounds	Substituted urea compounds

(*c*) *Other chemicals.* Acute toxicity data are available for approximately one hundred technical grade pesticides in general use, not listed in the above groups. These chemicals are either not likely to reach the marine environment, or, if used as directed by the registered label, probably would not occur at levels toxic to marine biota. It is presumed that criteria establisbed for these chemicals in fresh water will protect adequately the marine habitat. It should be emphasized that no unlisted chemical should be discharged into estuarine or coastal water without Preliminary bioassay and the establishment of an adequate application factor.

3. Other Toxicants

1. Allowable concentrations of metals, ammonia, cyanides and sulphides should be determined by use of the 96 h TL_m values and appropriate application factors. Preferably the TL_m values should be determined by flow-through bioassays in which environmental factors are maintained at least favourable levels. Tests should utilize the most sensitive life state of a species of ecological or economic importance in the area. Tentatively, it is suggested that application factors should be 0.01 for pesticides and metals, 0.05 for ammonia, 0.110 for cyanide and 0.05 for sulphides.
2. There is evidence that fluorides are accumulative in organisms. It is tentatively suggested that allowable levels should not exceed those for drinking water.

Comment. There are good grounds for asserting that the fluoride level in sea water is rapidly and effectively controlled by complex formation. What is said may be true for some drinking waters but the detailed chemistry must be thoroughly appraised. There is no evidence of accumulation of fluoride in seaweeds and grazing molluscs collected in the vicinity of a voluminous discharge containing a low level of fluoride.

3. The further dilution of wastes in marine waters suggests that the adoption of criteria established for detergents

and surfactants in fresh water (*i.e.*, the concentration of L.A.S. should not exceed 0.2 mg dm^{-3} or 0.14 of the 48 h TL_m concentration, which ever is the lower) also will protect adequately biota in the marine environment.

4. Bacteriological criteria of estuarine waters utilized for shellfish cultivation and harvesting should conform with the material standards provided that:

   (a) Examinations shall be conducted in accordance with recommended procedures for the examination of seawater and shellfish.

   (b) There shall be no direct discharges of untreated sewage within effective distance.

   (c) Samples of water for bacteriological examination shall be collected under those conditions of time and tide which produce maximum concentration of bacteria.

   (d) The coliform median M.P.N. of the water does not exceed 70/100 ml, and not more than 10 per cent of the samples ordinarily exceed an M.P.N. of 230/100 ml for a 5-tube decimal dilution test (or 330/100 ml where the 3-tube decimal dilution test is used) in those portions of the area most probably exposed to faecal contamination during the most unfavourable hydrographic and pollution conditions.

   (e) The reliability of nearby waste treatment plants should be considered in the approval of areas for direct harvesting.

5. It is also essential to monitor continuously waste from tar, gas and coke, petroleum refinery, petrochemical and pulp and paper mill operations. They all produce complex wastes of great variability, not only from facility to facility but also from day to day. This should be done on an individual basis with bioassays. These

tests should be made at frequent intervals to determine TL_m values as described for other wastes. For the more persistent toxicants, and application factor of 0.01 should be used while for unstable or biodegradable materials an application of 0.05 is tentatively suggested.

Comment. This statement puts the onus of monitoring on the control body; there are great advantages in having this done by the polluter with random checks by the controller. In many instances effluent analysis has encouraged the operator to reconsider losses to waste and economic benefit as well as effluent improvement have ensued.

6. Concentrations of other materials with noncumulative toxic effects should not exceed 0.10 of the 96 h TL_m value. For toxicants with cumulative effects, the concentrations should not exceed 0.05 if slowly degraded and 0.01 if stable of the above values.

7. *Multiple toxicants.* When two or more toxtc materials have additive effects in the receiving water, some reduction in the permissible concentrations as derived from bioassays on individual substances is necessary. The amount of reduction required is a function of both the number of toxic materials present and their concentrations with respect to the derived permissible concentration. An appropriate means of assuring that the combined amounts of the several substances do not exceed a permissible combination for the mixture is through the use of the following relationship:

$$\frac{C_a}{L_a}+\frac{C_b}{L_b}+\ldots+\frac{C_n}{L_n}\leqslant 1,$$

where C_a, C_b,..., C_n are the measured concentrations of the several toxic materials in the water and L_a, L_b..., L_n are the respective permissible concentrations (limits) derived for the materials on an individual basis. Should the sum of the several fractions exceed 1, then a local restriction on the concentration of one or more substances is necessary.

5

INDUSTRIAL WATER CONDITIONING

Industrial water conditioning is the mechanical and chemical treatment of natural and municipal water for efficient use in industrial processes. Industry uses enormous quantities of water as a heat-transfer medium for cooling products and equipment, to produce steam for electrical power generation, and as a raw material in the production process (Table 5.1).

TABLE 5.1

Water Usage by Industry[1]

Industry	*Water used per unit production*	*Water used per finished product*
Paper	20,000 gal/ton	1 gal/8 sheets of typing paper
Oil refining	20,000 gal/barrel crude	80 gal/gal gasoline
Steel	50,000 gal/ton	25 gal/1-lb box of nails
Power	360 gal/min MW	51 gal/100-W bulb burning for 24 hr

[1]Reprinted with permission from Hales, W.W. (1982). Use and reuse of water. *Chemtech 12*, 532. Copyright 1982 American Chemical Society.

The solvent property of water causes it to absorb practically any substance it comes into contact with. Since these sub-stances can corrode, form deposits, and foul system metallurgy, it is extremely important to properly condition the water prior to its use in industrial processes. This article focuses on the principles involved in the conditioning of water for use in steam generation and for the removal and transportation of excess heat, which account for approximately 90 per cent of all industrial water use.

I. WATER: NATURE AND CONTENT

A. Hydrologic Cycle

The chief source of water is moisture that vapourizes from land masses and oceans and subsequently condenses in the atmos-phere to precipitate as rain, snow, sleet, or hail. This is known as the hydrologic cycle. As it descends through the atmosphere, water is contaminated by atmospheric gases such as oxygen, carbon dioxide, and sulphur dioxide. Solids in the atmosphere, such as pollen, spores, dust particles, smoke, and fumes, also contribute impurities to the condensed water vapour.

Abundant supplies of fresh water are essential to industrial development. Since its availability is limited, it is extremely important to properly condition and conserve our fresh-water supplies. [See *Water Resources Systems*.]

Surface water and groundwater, the two most important sources of fresh water, have some singificant differences in their physical characteristics. Surface water collects in ponds or lakes or runs off in streams and rivers. This water accumulates silt and minerals as it flows across the terrain. The kind and quantity of dissolved and suspended solids in the flowing streams depend on the geological formations in the particular area. Surface waters can also contain various macro-and microscopic life forms, which present additional considerations in the water treatment process. Additionally, riverine sources may change quality characteristics in a short period of time.

Groundwater collects in pockets or underground steams and forms sources for wells and springs. Like surface water, the nature of the impurities found in groundwater depends upon the geographical region in which it is located. Due to the decay of organic matter in the earth, groundwater normally has a higher corbon dioxide content than surface water. The use of groundwater is advantageous since it is usually free of suspended matter and is constant in quality and temperature. However, dissolved gases and mineral content are often high.

Salt water from the oceans or bays is also an important source of water for industrial purposes. In those industries located near coastlines, saltwater is used extensively in cooling systems. Saltwater is also processed for domestic use.

B. Conlent: Solids and Gases

Impurities that are commonly found in water can be broadly classified as dissolved solids, suspended solids, and gases (see Table 5.2). Even very minute quantities of these impurities (as measured in milligrams per litre or parts per million) can be significant in water conditioning. Precise analytical tests are required to quantify the amounts and types of impurities, as well as the results of treatment.

II. PROCESSING OF WATER

A. Clarification

Clarification is the removal of suspended matter from raw water supplies to make water more suitable for industrial or domestic requirements. The process involves the coagulation, flocculation, and sedimentation of particulates. [See *Westewater Treatment.*]

Colloidal suspended particles in water have like electrical charges that tend to keep them in permanent suspension. Coagulation is a process of charge neutralization so that suspended particles no longer repel each other. Flocculation involves the collection of these neutralized particles in clumps called floc. The third part of the process, sedimentation, is the settling or removal of the floc from water. Sedimentation alone, without coagulation, results in the removal of coarse

TABLE 5.2

Impurities in Water

Substance	*Formula or definition*	*Possible problems*	*Possible treatment*
Alkalinity	Bicarbonate (HCO_3^-), carbonate (CO_3^{2-}), and hydrate (OH^-), expressed as $CaCO_3$	Foaming and carryover in boilers; corrosion in condensate lines; embrittlement of boiler steel	Like and lime-soda softening; hydrogen zeolite softening; demineralization; dealkalization
Carbon dioxide	CO_2	Corrosion in water lines and particularly steam and condensate lines	Aeration; deaeration; neutralization with alkalies
Chloride	Cl^-	Addition to solids content and increase in corrosive character of water	Demineralization
Hardness	Primarily calcium and magnesium ions expressed as $CaCO_3$	Scale in heat exchange equipment, boilers, pipelines, and processes	Softening; demineralization; internal boiler water treatment; surface-active agents
Iron	Fe^{2+} (ferrous), Fe^{3+} (ferric)	Discolouring of water on precipitation; source of deposits in water lines,	Aeration; coagulation and filtration; lime softening cation exchange; contact

		boilers, etc.; interference with dyeing, tanning, papermaking, etc.	filtration; surface-active agents for iron retention
Oxygen	O_2	Corrosion of water lines, heat exchange equipment, boilers, return lines, etc.	Deaeration; sodium sulphite; corrosion inhibitors
pH	Hydrogen ion concentration defined as follows: pH= log(1[H⁺] where [H+] is molar concentration	pH variance according to acidic or alkaline solids in water; most natural waters of pH of 6.0—8.0	pH increased by alkalies and decreased by acids
Silica	SiO_2	Scale in boilers and cooling water systems; insoluble turbine blade deposits due to silica vapourization	Hot process removal with magnesium salts; adsorption by highly basic anion exchange resins; demineralization
Sulphate	SO_4^{2-}	Addition to solids content of water; combination with calcium to form calcium sulphate scale	Demineralization
Turbidity	Expressed in analysis as units	Unsightly appearance imparted to water; deposits in water lines, process equipment, etc.; interference with most process uses	Coagulation, settling, and filtration

suspended solids but does not usually produce water sufficiently clarified for efficient industrial use.

Clarification processes are also employed for removal of colour that might be present in water supplies. Colour-producing compounds are organic acids that are picked up by water in runoff over geologic formations. Most organic colour in surface waters is colloidal and negatively charged.

1. Coagulation

Inorganic salts of aluminium or iron are primary coagulants that neutralize the charges on suspended particles. The salts also hydrolyze to form insoluble precipitates that entrap additional particles. Flocculation begins after the addition of the primary coagulant as the neutralized particles begin colliding and growing in size. This process. may occur naturally, or the speed of reaction can be increased by the addition of inorganic or polymeric coagulant aids. A list of common coagulants and coagulant aids is given in Table 5.3.

TABLE 5.3

Common Coagulants and Coagulant Aids

Name	*Formula*	*Commercial strength*
Aluminium sulphate	$Al_2(So_4)_3 \cdot 18H_2O$	17% Al_2O_3
Sodium aluminate	$Na_2Al_2O_4$	55% Al_2O_3
Ammonium alum	$Al_2(SO_4)_3 \cdot (NH_4)_2SO_4 \cdot 24H_2O$	11% Al_2O_3
Copperas	$FeSO_4 \cdot 7H_2O$	55% $FeSO_4$
Ferric sulphate	$Fe_2(SO_4)_3$	90% $Fe_2(SO_4)_3$
Ferric chloride	$FeCl_3 \cdot 6H_2O$	60% $FeCl_3$
Magnesium oxide	MgO	95% MgO
Bentonite	—	—
Sodium silicate	$NA_2 \cdot 3.2SiO_2$	—

An important parameter in the clarification process is the pH of the water. With aluminium sulphate, the optimum coagulation efficiency normally occurs in the range of 6.0 to 7.0. Sodium aluminate is highly alkaline and is used at elevated pH. Iron coagulants can be used successfully over the much broader pH range of 5.0 to 11.0.

2. Polyelectrolytes

Polyelectrolytes are water-soluble organic polymers used in the clarification process as primary coagulants or coagulant aids. Polymers are classified as anionic, cationic, or nonionic. Anionic polymers ionize in water solution to form negative sites along the polymer molecule. Cationic polymers ionize to form positively charged sites, while nonionic polymers exhibit only slight ionization.

Polymeric primary coagulants are cationic polyamines with very high cationic charge density. They function by adsorbing onto particle surfaces and neutralizing the negative charges.

Polymeric coagulant aids can be used with either inorganic coagulants or polymeric coagulants. These aids may be anionic, cationic, or approaching neutral charge and are usually of extremely high molecular weight. They promote flocculation primarily be bridging particles together.

Since coagulation is an art rather than a science, specific testing of a turbid water supply is necessary to determine the optimum combination of polymeric or inorganic coagulants and flocculants.

3. Types of Equipments

Three separate unit processes are involved in conventional clarification:

(1) rapid mix for coagulation,

(2) slow mix for flocculation, and

(3) floc-water separation.

The oldest, most basic type of system is a horizontal flow-through unit. This unit is usually divided into two or three sections. Chemical feed is introduced in a rapid mix area, which is followed by slow mix and settling sections. Rapid mix is typically designed for 3-to 5-min duration and slow mix for 15 to 30 min, Retention time for horizontal units is long (in the area of 4 to 6 hr) primarily to allow for subsidence of the floc.

Today, industrial clarification is more commonly accomplished in compact, circular steel or concrete units of upflow design. The term upflow describes this unit's sedimentation process, where the water flow is directed upward while the suspended solids settle downward. Clarifiers of the upflow type are also referred to as solids-contact units because large volumes of sludge are circulated internally. Some solids-contact units are referred to as sludge blanket clarifiers, but essentially the same mechanism is employed. The efficiency of the unit depends on the filtering action as the freshly coagulated or flocculated water passes through the suspended floc. Higher sludge levels increase the filtration efficiency. With good operation, a distinct sludge-liquid interface is formed. A high-clarify water is obtained as a result of the increased solids contact through internal sludge recirculation.

An important parameter in the operation of an upflow unit is the rise rate, which is the upward flow of water through a unit of surface area. The normal rise rate for clarification is in the range of 0.7 to 1.25 gal/min per square foot of surface area (28-50 litres/min m^2). The holding capacity (retention time) of an upflow unit is 1 to 2 hr, making it much more efficient in space requirement than a horizontal unit of the same throughput.

4. In-Line Clarification

The removal of turbidity from raw water can often be accomplished by adding coagulants to the water just prior to the filtration process. Either inorganic coagulants or polymers can be used, but the polymers are advantageous in that they do not place additional suspended solid loading on the filters. The use of a cationic polymer coagulant is selected on the basis of operating experience or specific testing on the water being treated. Complete charge neutralization and bridging are not necessary, and so the feed-rate is usually less than that for conventional clarification.

5. Jar Testing

A multiple-paddle beaker arrangement, referred to as a jar tester, is used to select optimum coagulation chemistry and mechanical conditions. The jar tester is a valuable asset in setting up a new clarification programme or in making modifications to an existing programme. Raw water characteristics frequently change due to seasonal and climatic conditions, and consequently require compensatory treatment modifications.

B. Filtration

1. Types of Filter Media

Filtration is employed where water is to be used for domestic, boiler makeup, or special process purposes. Water can be filtered by passing it through a bed of porous media such as silica sand, quartz sand, garnet, magnetite, anthracite coal, or a combination of these. The turbidity of water, which is an indirect measure of suspended solid content, can be significantly reduced by conventional sand filtration.

The filter medium gradully becomes fouled with the particulate matter it removes from the water. After a predetermined service run, usually based on head loss, the filter is "backwashed" to remove the particulate matter and ready the filter for the following service run.

Different water flow patterns and a variety of filtration media and particle sizes can be used to design a filtering system to satisfy the specifications of an individual plant process.

2. Types of Filters

Filters may be divided into two basic types of operation;

gravity and pressure. The oldest type is the gravity filter in which water simply flows downward through the medium. A disadvantage of the gravity type is that the filtered water must usually be repumped to its destination.

In pressure-type filters, the medium is enclosed in a vertically or horizontally aligned vessel so that a pumping head can be applied to force the water through the filtration medium.

The service flow rate of pressure filters is usually 3 gpm/ft^2 of surface area. Dual or multimedia filters are designed for 6 gmp/ft and higher. The backwash rate for pressure filters depends on the temerature and filter medium. Lower flow rates are required at ambient temperature because of the greater density of the backwash water. The backwash rate for an anthracite medium is 6 to 8 gpm/ft^2 as compared to 13 to 15 gpm/ft^2 required for sand.

Multimedia and in-depth filtere combine media of different particle sizes and densities so that high water flow rates or longer filter runs can be obtained. Particulate matter is removed from the water in a more uniform distribution pattern throughout the bed.

3. Precoat Filters

Where relatively small quantities of water require final polishing for the removal of contaminants, a system of precoat filtration can be employed. Extremely fine particulate matter, oil, and even some microorganisms can be removed with this specialized type of filtration.

Water is forced under pressure through a precoat of diatomaceous earth and a permeable base. Porous stone tubes, wire screens, porous papers, and wire-wound tubes are used as base materials. The precoat is applied on a cyclical basis. The base material permits passage of filtered water while holding the precoat in position.

C. Aeration

1. Equipment

Aeration is a process used for stripping gases and oxidizing dissolved metals.

Two general methods are used for industrial aeration. The more common is forced draft aeration, which employs equipment that pumps air through the water. Water is broken up into small droplets or a thin film to produce maximum surface area contact with the countercurrent air flow.

The second method is air diffusion, which is accomplished by pumping air into water through perforated pipes, strainers, porous plates, or tubes. Aeration by diffusion is theoretically superior to waterfall aeration, since a fine bubble of air rising through water is continually exposed to fresh liquid surfaces.

2. *Dissolved Gases*

Aeration is used to reduce the concentration of carbon dioxide (decarbonation), ammonia, and hydrogen suphide in water. Unlike oxygen, these gases ionize in water. To remove carbon dioxide and hydrogen sulphide, the pH of the water is lowered by adding acid.

For example, over 95% of carbon dioxide can be removed from ambient water at pH of 5.0. At a pH of 7.0, only 20% of CO_2 is removable. Ammonia removal efficiency can be improved by increasing pH through the addition of caustic soda. Over 95% of ammonia can be removed at a pH of 11.0.

Removal of gases by aeration is enhanced by higher temperatures, longer aeration time, and larger volumes of air in contact with the water. However, aeration also results in the saturation of water with dissolved oxygen, which is also a corrosive gas.

3. *Iron and Manganese Removal*

Iron and manganese are present in many well water supplies in the soluble ferrous and manganous bicarbonate forms. In the process of aeration these metals are oxidized to the insoluble higher-valence forms ferric and manganic, which subsequently can be removed by settling or filtration. Manganese is more difficult to remove than iron and usually requirtes elevation of pH for removal. Oxidants such as chlorine or potassium permanganate are sometimes added to ensure complete oxidation.

A modification of the forced draft aerator with trays of coke is often used for the iron and manganese oxidation processes. Contact of the iron-manganese-bearing water with fresh carbon surfaces catalyzes the oxidation reaction.

D. Chlorination

1. Uses and Effects

Chlorine is commercially available in liquid and solid forms and is widely used in water treatment. Compounds of chlorine such as hypochlorites, chlorine dióxide, and organic chlorine donors also have varying degrees of usefulness. Some of the important chlorine applications are (1) disinfection through biocidal activity, (2) control of taste, colour, and odour, (3) removal of ammonia and other nitrogeneous organics, (4) hydrogen sulphide oxidation, and (5\ iron and manganese removal. The prime oxidizing agent when chlorine is dissolved in water is hypochlorous acid. Reactions are

$$\underset{\text{chlorine}}{Cl_2} + \underset{\text{water}}{H_2O} \Rightarrow \underset{\text{hypochlorous acid}}{HOCl} + \underset{\text{hydrochloric acid}}{HCl} \quad (1)$$

$$\underset{\text{hypochlorous acid}}{HOCl} \leftrightarrows \underset{\text{hydrogen ion}}{H^+} + \underset{\text{hypychlorite ion}}{OCl^-} \quad (2)$$

The pH of the water is a factor in chlorination. At lower pH. hypochlorous acid (the stronger oxidizing species) predominates in accordance with Eq. (2). The net effect is that chlorination is more efficient when the pH of the water is below 7.0 than it is above 7.0.

Industrial water treatment usually involves the disinfection of water rather than its sterilization. Disinfection implies the deactivation of pathogenic organisms and the control of growth of other organisms. For example, a free chlorine residual of 0.3 to 0.6 ppm is usually sufficient to control bacterial slime and algae growth in cooling water systems.

Chlorine is so highly reactive and effective at low concentrations that precise analytical techniques are required. Analyses are used to differentiate types of chlorine residuals that exist in a water supply and their efficiency.

1. Free chlorine residual is unreacted hypochlorous acid and hypochlorite ion.
2. Combined chlorine residual is the chlorine that exists in water in chemical combination with ammonia or other organic nitrogen compounds.
3. Total chlorine residual is the total amount of free and combined chlorine existing in water.

The difference between the amount of chlorine added to water and the amount of free, combined, or total residual that remains after a specific contact time is called the chlorine demand of a system.

2. Hypochlorination

Commercial forms of hypochlorite are widely used for disinfection. Sodium hypochlorite (NaOCl) is available in several concentrations, ranging from common household bleach with 5 per cent available chlorine through commercial laundry bleaches that contain 12 per cent—15 per cent available chlorine. Bleach solutions contain an excess of alkali that elevates the pH during application of the bleach. This tends to reduce the efficiency of the chlorination and can result in precipitation of calcium carbonate. On the other hand, the high alkalinity of hypochlorite treatment improves the general condition of soft and highly corrosive waters.

Another commercial form of hypochlorite is calcium hypochlorite, which contains 7 per cent available chlorine. Calcium hypochlorite is soluble in water but often forms a precipitate in hard water supplies.

3. Chlorine Dioxide

Chlorine dioxide (ClO_2) is used commercially for bleaching and disinfecting various process waters. It has 2.5 times the oxidizing power of chlorine. Chlorine dioxide does not dissociate in water but forms a true solution of gas in water. Therefore, the pH of the water does not affect the efficiency of the application.

Chlorine dioxide is extremely unstable and is usually generated at the point of use. The principal reaction for formation of chlorine dioxide is between chlorine and sodium chlorite.

Chlorine dioxide is much more expensive than chlorine; some advantages of chlorine dioxide are (1) that it does not react with ammonia and (2) that its stronger oxidizing power is desirable for some organic materials. Treatment of process effluents with chlorine dioxide will destroy phenols without creating the taste problem associated with chlorinated phenols.

4. Chlorine Donors

Some organic chlorine compounds such as cyanurates and hydantoins hydrolyze in water to provide chlorine residuals. These materials are advantageous in swimming pool treatment and in some small industrial water treatment applications where ease of handling is desired. The cyanurates also protect the chlorine from the destructive effects of UV light.

5. Dechlorination

The removal of chlorine from water is often necessary or desirable for further plant use or to discharge the water to the environment. Chlorine residuals can adversely affect industrial processes and materials such as ion exchange resins.

One common method of dechlorination is to pass the water through a bed of activated carbon. Chemical dechlorination with sulphur dioxide, sodium sulphite, or sodium bisulphite is also practiced.

E. Precipitation Processes

1. Chemistry

Many natural water supplies require treatment to reduce hardness, alkalinity, silica, and other constituents to satisfy the demands of the process. Boiler makeup water and cooling water specifications often necessitate this type of treatment on a water supply. precipitation processes performed at ambient or elevated temperatures are employed to fulfill the needs of the process.

Treatment is accomplished by (1) lime, or combinations of lime, with (2) soda ash, (3) gypsum, and (4) sodium aluminate. These chemicals react with hardness and natural alkalinity in the water to form insoluble compounds that precipitate and are removed from the water by sedimentation.

Cold lime softening is the descriptive term applied to the process in which hydrated lime, $Ca(OH)_2$, is applied at ambient temperature. By maintaining the proper chemical control and lime feed, the calcium hardness may be reduced to 35 to 50 ppm. Magnesium reduction is a function of the amount of hydroxyl (OH^-) alkalinity excess maintained. Figure 5.1 shows the solubility of calcium and magnesium in the processes.

Some of the principal reactions in cold lime softening are

$$\underset{\text{carbon dioxide}}{CO_2} + \underset{\text{calcium hydroxide}}{Ca(OH)_2} \longrightarrow \underset{\text{calcium carbonate}}{CaCO_3} + \underset{\text{water}}{H_2O} \tag{3}$$

$$\underset{\text{calcium bicarbonate}}{Ca(HCO_3)_2} + \underset{\text{calcium hydroxide}}{Ca(OH)_2} \longrightarrow \underset{\text{calcium carbonate}}{2CaCO_3} + \underset{\text{water}}{2H_2O} \tag{4}$$

$$\underset{\text{magnesium bicarbonate}}{Mg(HCO_3)_2} + \underset{\text{calcium hydroxide}}{2Ca(OH)_2} \longrightarrow \underset{\text{magnesium hydroxide}}{Mg(OH)_2} + \underset{\text{calcium carbonate}}{2CaCO_3} + 2H_2O \tag{5}$$

Soda ash (Na_2CO_3) is used with lime to improve hardness reduction where the calcium and magnesium are present as noncarbonate hardness. Gypsum ($CaSO_4.2H_2O$) is added in some processing where the alkalinity of the raw water is higher

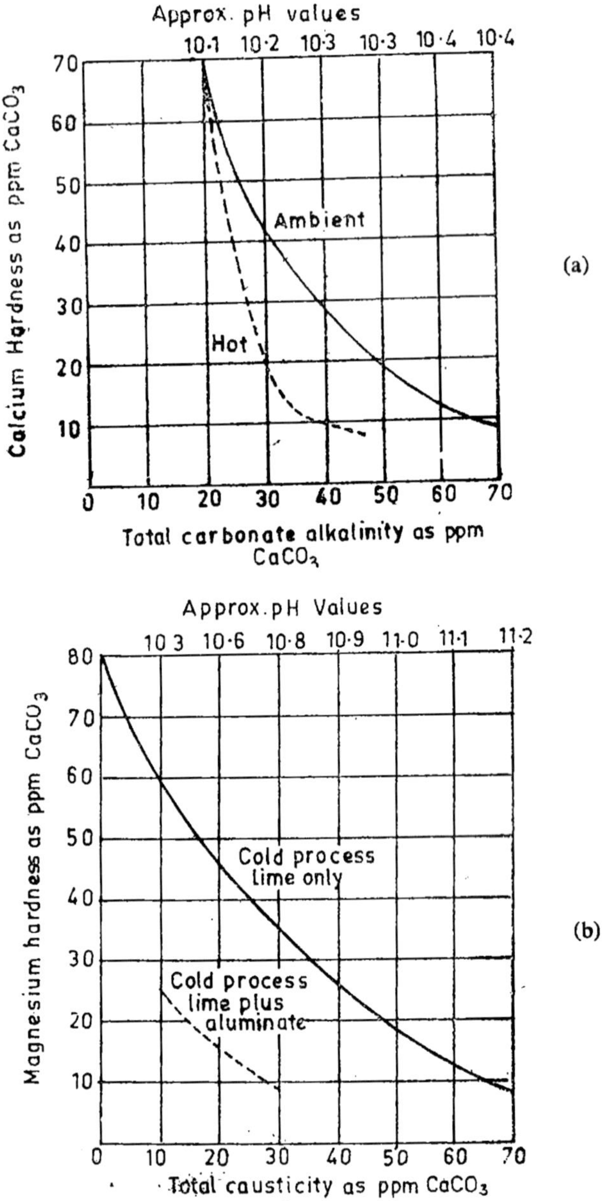

Fig. 5.1: (a) Calcium reduction versus carbonate alkalinity
(b) Magnesium reduction versus causticity

than the hardness. In these cases, gypsum provides the calcium for precipitation of calcium carbonate.

Sodium aluminate, $Na_2Al_2O_4.12H_2O$, is used to improve magnesium reduction, which, in turn, improves silica removal.

2. Cold-Warm Processes

Most softening processes are carried out at ambient temperature with equipment similar to that used for the clarification process. If the temperature of the process is raised to the range of 120 to 140°F, the process is termed warm lime softening. While reaction rates are improved at the higher temperature, the control over temperature becomes very critical. Variations of as little as 3°F to 5°F per hour can cause gross carryover.

Water flow rate in a typical unit is 1.5 gpm ft^2 of area. Retention time of 1.5 hr usually permits fairly complete reaction time. However, the resulting effluent is unstable and further precipitation downstream is possible. Acid or carbon dioxide is added to reduce the pH from the area of 10.2 to between 7.5 and 8.0 in order to stabilize the stream.

Control over softener operation is usually based on treated water alkalinity and hardness. Tests are conducted on the outgoing stream to determine the alkalinity to the phenolphthalein (P) (pH 8.3) and the methyl orange (M) (pH 4.3) end points. When the methyl orange alkalinity is less than twice the phenolphthalein alkalinity. the hydroxyl (OH^-) ion is present for good softener operation. Therefore, the usual control is to maintain the 2P—M value in the range of +5 to +15 ppm. This corresponds to a pH of approximately 10.2.

Where soda ash is used, an additional control parameter is typically maintained as follows:

$$\text{M (alkalinity)} - \text{TH (total hardness)} = +20 \text{ to } +40$$

3. Hot Process

When a pressure vessel is used and the temperature is in the range of 227°F to 240°F, the plant is practicing hot process softening. The chemistry is similar to that of the cold-warm processes, but the reactions at these elevated temperatures are more complete. Equipment is also available for downflow operation.

A hot process unit will reduce oxygen to about 0.2 cm^3/litre and carbon dioxide to 0.0 cm^3/litre as a result of heating the water. A comparison of cold and hot process softening is given in Table 5.4. Both cold and hot precipitation processes are also effective in removing iron, manganese, turbidity, and some organics, including colour bodies.

F. Ion Exchange

1. Materials

Ion exchange materials are widely used to treat raw water supplies that contain dissolved salts. These materials have the unique ability to exchange one ion for another, hold it temporarily, and release it to a regenerant solution. The proper regenerant can replace undesirable ions taken from the water by the exchange material with a fresh supply of acceptable ions.

For example, a sodium zeolite softener replaces scale-forming calcium and magnesium ions in the water with sodium ions. One of the original processes for softening water used sodium aluminosilicate, commonly called zeolite. Today, efficient, high-capacity synthetic resins have replaced this natural material, but the term zeolite is still used to describe any cation exchange process.

Polystyrene divinyl benezene and carboxylic-based resins are the primary materials used in water treatment. The resins are further classified as having either a gelular, permeable membranetype structure or a macroreticular-(macroporous)-type structure. Macroreticular cationic resins have discrete pores that provide high resistance to thermal shock and oxidation. Macroreticular anionic resins resist organic fouling and are commonly used upstream of a gelular resin to protect it from raw water organics.

TABLE 5.4

Typical Softener Effluent Analysis

	Raw water	*Cold-lime*	*Lime-soda softening (cold)*	*Lime-soda softening (hot)*	*Lime softening (hot)*
Total hardness (as $CaCO_3$), ppm	250	95	81	20	107
Calcium hardness (as $CaCO_3$), ppm	150	35	35	15	102
Magnesium hardness (as $CaCO_3$), ppm	100	60	46	5	5
P alkalinity (as $CaCO_3$), ppm	0	27	37	23	18
M alkalinity (as $CaCO_3$), ppm	150	44	55	40	25
Silica (as SiO_2), ppm	20	19	18	1-2	1-2
pH	7.5	10.3	10 6	10.5	10.4

Water treatment resins are also classified as one of four basic process types:

1. strong cation (SC),
2. weak cation (WC),
3. strong-base anion (SB), and
4. weak-base anion (WB).

Strong-base anion exchange resins are also commonly referred to as type I and type II resins. Type I has greater chemical stability, while type II has a slightly greater regeneration efficiency and capacity.

2. *Zeolite Softening*

Sodium zeolite softening, the most widely practiced form of ion exchange, prepares water for use as boiler feedwater and in numerous other industrial processes. The zeolite softening process exchanges hardness ions of calcium and magnesium for sodium ions. Strong cation exchange resins of the polystyrene type are used to effect the exchange.

Basically, this is a four-step process that consists of (1) service run, (2) backwash, (3) regeneration, and (4) rinsing. An efficient flow rate for the service run is 6-8 gpm/ft^2 of resin area. The length of the service run depends on the capacity of the resin and the hardness of the water.

During backwashing, the flow is in the reverse direction of the service run, which expands the bed and releases fouling matter. The backwash rate is specified by the manufacturer and depends on temperature and resin characteristics. The rate is intended to obtain a 50 per cent expansion of the bed for 10 min.

Regeneration is accomplished with a 10 per cent salt solution at a rate of approximately 0.5 to 1.0 gpm/ft^3 of resin. Regeneration displaces the hardness that is collected by the resin during the service run with fresh sodium to soften the water during the succeeding run. The capacity of the resin varies with the amount of salt added. Typically, 8 lb of salt per cubic foot of resin provides a capacity of 20,000 grains of hardness exchange per cubic foot.

Rinsing completes the regeneration process by displacing any excess salt solution. After a short period of slow rinse to ensure adequate adsoption of the regenerant, a fast rinse at a rate of 1.5 gpm/ft^3 of resin is applied until the salt is removed from the unit.

Sodium zeolite softening is operated at elevated temperatures (227-240°F) following a hot lime precipitation process in many industrial plants. Polystyrene resin has good operating efficiency at the elevated temperature, and the combination treatment provides good quality makeup water for medium-pressure boiler systems.

3. *Demineralization*

Higher-purity water is required for numerous industrial uses such as high-pressure boiler feedwater, electronics production, pharmaceutical manufacturing, and chemical processing. Demineralization can produce water of a purity comparable to distilled water (*i.e.*, 1 μmho conductivity) and at a much lower cost. Demineralization removes essentially all inorganic salts by ion exchange. The process consists of hydrogen cation exchange, which converts dissolved salts to their corresponding acids, followed by basic anion exchange, which removes the acids. Demineralizer systems are operated in many different sizes and configurations.

The cation units of demineralizer systems are operated on the hydrogen cycle with acid as the regenerant. Highly alkaline raw water may require a decarbonator or vacuum deaerator for economical removal of carbon dioxide.

Silica removal requires the use of strong-base anion resins. A 4 per cent caustic soda solution applied at a temperature of approximately 120°F is used for efficient regeneration. Where silica removal is not critical, weak-base anion resins are used. Caustic soda, ammonia, or soda ash may be used for regeneration of weak-base anion exchange resins. The service flow rate of demineralizers is usually in the range of 6 to 8 gpm/ft^2 of resin surface.

In mixed bed demineralizers that are used for producing water of extremely high purity, strongly acidic cation and strongly basic anion resins are mixed for the service run. The resins are separated for regeneration.

4. Other Processes

Hydrogen zeolite softening removes calcium, magnesium, and sodium ions from water, replacing them with hydrogen ions. Sulphuric acid is the usual regenerant, although other acids may be used. This process is used as the first step in demineralization or for partial processing of raw water with high alkalinity.

Three processes that are used for softening and removing alkalinity (dealkalization) from the raw water are (1) sodium zeolite-hydrogen zeolite split stream softening, (2) sodium zeolite-chloride anion dealkalization, and (3) weakly acidic cation exchange.

Most ion exchange units operate in a batch mode, as described. However, numerous commercial applications of continuous ion exchange processes exist.

5. Condensate Polishing

Ion exchange is used extensively for polishing condensate from boiler systems for reuse as boiler feedwater. Strong cation resins stable at temperatures up to 270°F and operating in the sodium zeolite cycle are typically used for polishing condensate. Operation of these units at flow rates in the range of 35 to 50 gpm/ft^2 removes undesirable condensate contaminants such as iron and copper.

In plant operations require the removal of all ions from condensate, a mixed bed resin column is used and temperatures are maintained below 140°F for the exchange process.

G. Deaeration

1. Mechanism

Deaeration is used to remove dissolved corrosive gases from

plant water streams. The most important application is the removal of oxygen from boiler feedwater systems.

A dissolved gas can be removed from water by reducing the partial pressure of that gas in the atmosphere contacting the liquid. In preboiler systems this is accomplished by applying steam as a purge and venting to remove the undesired gas. The solubility of dissolved oxygen in water as a function of temperature is given in Fig. 5.2. Operation of a deaerating heater to within a few degrees of the temperature of saturated steam results in removal of 97 per cent to 98 per cent of the oxygen.

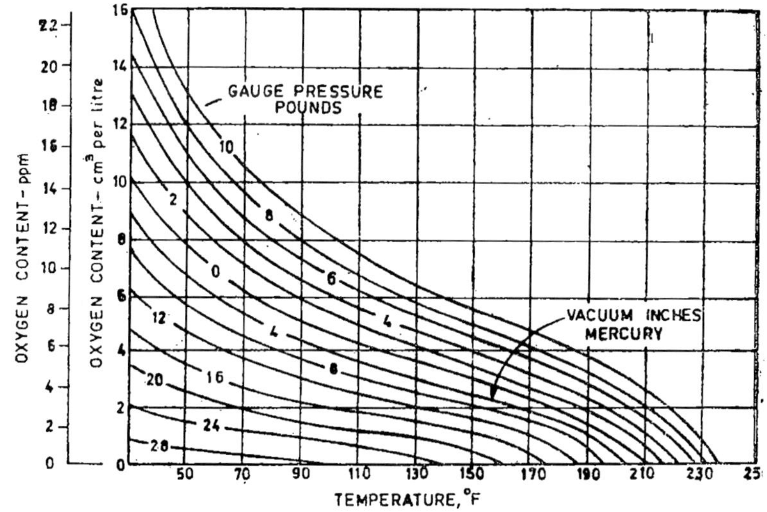

Fig. 5.2: Solubility of dissolved oxygen

2. Equipment

The two major types of pressure deaeratars or deaerating heaters are the tray type and the spray type.

Deaerating heaters are operated as close to saturation conditions as possible; for example, 227°F at 20 lb/in.^{2} absolute pressure. Under these conditions, heaters usually produce water with less than 0.005 cm^3/litre of oxygen and 0.000 free carbon dioxide. Since oxygen is extremely corrosive to a boiler system, the trace amounts that remain following mechanical deaeration are usually removed with a chemical oxygen scavenger.

III. BOILER WATER TREATMENT

A. Corrosion Control

Boiler metal corrosion can be caused by dissolved gases, chemicals, concentration cells, and low pH. Control over corrosion is complicated by the elevated temperature, pressure, and mixed metallurgies represented in an operating boiler system.

1. Oxygen Pitting

Oxygen pitting is one of the most common and serious of the possible types of attack on boiler and preboiler metallurgy. Even traces of oxygen left in the feedwater can cause failure of boiler systems. Mechanical deaeration of the feedwater is an important first step in eliminating dissolved oxygen and other corrosive gases such as ammonia, carbon dioxide, and hydrogen sulphide. For complete oxygen removal, chemical assistance is necessary. Where mechanical deaeration is not available, chemical oxygen scavenging alone can sometimes provide adequate protection. Economics generally favour a combination of both mechanical and chemical oxygen removal.

The most common oxygen scavenger is sodium sulphite, which combines with oxygen according to the reaction

$$\underset{\text{sodium sulphite}}{2Na_2SO_3} + \underset{\text{oxygen}}{O_2} \longrightarrow \underset{\text{sodium sulphate}}{2Na_2SO_4}$$

The normal sulphite control range in boiler water is 20-60 ppm as SO_3. The use of sodium sulphite for oxygen control is generally limited to boilers operating below 900 lb/in.^{2} because sulphite can decompose at higher pressures. The products of decomposition are sulphur dioxide and hydrogen sulphide, both of which can cause corrosion.

The normal feed rate for sodium sulphite is 10 lb for each pound of oxygen present in the feedwater. Continuous feed of the sulphite corrective is desirable, and the usual feed point is the storage section of the heater.

The speed of the reaction between oxygen and the chemical deaerant is an important consideration in selection of a specific agent. The reaction of sulphite with oxygen can be increased in rate by the use of a catalyst, by temperatures in excess of 212°F and a pH of 9.0 to 10.0. Effective catalysts are heavy metal cations such as iron, cobalt, and nickel. Catalyzed sodium sulphite is commonly used at 5 to 15 ppm in the feedwater to ensure instantaneous removal of oxygen in a preboiler system, particularly where an economizer is involved. Economizers transfer heat from the flue gas to the feedwater.

Hydrazine (N_2H_4) is used extensively as an oxygen scavenging agent in boiler systems operating at 900 lb/in.^{2} and above. Hydrazine reacts according to the equation

$$\underset{\text{(hydrazine)}}{N_2H_4} + \underset{\text{(oxygen)}}{O_2} = \underset{\text{(water)}}{2H_2O} + \underset{\text{(nitrogen)}}{N_2}$$

Hydrazine adds no solids to the boiler water, an important consideration in the operation of high-pressure boilers. Hydrazine also promotes the formation of the protective magnetite (Fe_3O_4) film.

Normally, 1.5-2.0 parts of hydrazine are required to compensate for each part of oxygen. Control of hydrazine feed is normally based on maintaining a feedwater excess concentration of between 0.05 and 0.1 ppm.

In the past few years, water treatment research has led to the commercial development of several organic agents as oxygen scavenging compounds. The research was motivated by the need for a hydrazine substitute, since hydrazine is toxic and a suspected animal carcinogen. Included in this development are oxygen scavengers based on hydroquinone, carbohydrazide, diethylhydroxylamine, and ascorbic acid.

2. *Acid and Caustic Attack*

Corrosion caused by low-pH boiler water can also be serious. Normally, boilers are protected by a magnetite film, which is produced by the maintenance of reducing conditions and a high-pH water chemistry.

In high-pressure boiler operations using demineralized makeup water, the potential for high-pH (caustic) attack requires specialized treatment.

The principal anion present in relatively pure, high-pressure boiler water is hydroxide. If this anion is concentrated, it can destroy the protective magnetite film and dissolve the base metal. Highly caustic anion concentrations can be formed under porous iron oxide deposits resulting in severe corrosion of tube surfaces with high heat flux.

Protection against caustic attack is obtained by the maintenance of a coordinated pH/phosphate boiler water chemistry programme. Disodium phosphate is used as a pH buffer; it combines with excess caustic to form trisodium phosphate. In some systems control of the molar ratio of sodium to phosphate (*e.g.*, 2.8/1.0) is maintained to prevent the formation of free caustic.

Low pressure steam and hot-water heating boilers requiring little or no makeup are generally protected by film-forming inhibitors of chromate or nitrite. Residuals of 800 to 1200 ppm of chromate or nitrite are usually sufficient to maintain a protective iron oxide film.

3. *Boiler Storage*

In certain operations and geographical areas, boilers are stored for various lengths of time. Idle boilers must be properly stored to prevent severe corrosion. Two methods of storage, dry and wet, are utilized. As the name implies, dry storage requires complete dehumidification of the boiler's internal surfaces. This is usually accomplished by maintaining a chemical desiccant in the equipment.

Proper wet storage of a boiler is obtained by completely filling the boiler to normal water level with deaerated feedwater. Three pounds of caustic soda and 1.5 lb of sodium sulphite are added per 1000 gallons of water (to obtain a minimum of 100 ppm phenolthalein alkalinity and 100 ppm of sulphite). An alternative treatment consists of 4.8 lb of a 35 per cent solution of hydrazine and 0.1 lb of ammonia per 1000 gallons of water (for a minimum 200 ppm N_2H_4 and 10.0 pH).

4. Embrittlement

Another form of corrosion-induced failure of boilers is intercrystalline cracking caused by caustic embrittlement of metal. This failure develops in the presence of a highly concentrated caustic solution when the metal is highly stressed. With no apparent signs of prior attack, cracking can suddenly occur along the grain boundaries of the crystal structure of the metal.

An embrittlement detector (developed by the U.S. Bureau of Mines) installed in the continuous blowdown line of the boiler can be used to determine whether the boiler water contains embrittling tendencies.

In low-pressure boilers, sodium nitrate is standard treatment for inhibiting embrittlement. Inhibition of embrittlement requires the proper ratio of nitrate to the caustic alkalinity present in the boiler water.

Coordinated phosphate/pH control treatments for caustic corrosion are also applicable for preventing caustic embrittlement of boilers.

B. Deposition Control

1. Types of Deposits

Deposition of boiler water solids on heat transfer surfaces can result in overheating and possible failure of boiler tubes. Deposits tend to form in the highest heat flux areas of the boiler. Figure 5.3 illustrates the insulating effect of deposition, which elevates tube wall temperatures.

Most industrial boiler operations produce good-quality feedwater by external treatment though zeolite softening or demineralization processes, averaging from 0.01 to 2.0 ppm total hardness. Additional internal boiler water treatment is still necessary for the prevention of harmful deposits.

The solubility of hardness salts in water decreases as temperature increases. These inversely soluble salts (*e.g.*, carbonate and silicate) precipitate from solution at higher temperatures and form crystalline deposits. Heat also breaks down the natural bicarbonate alkalinity of the feedwater to form carbon dioxide (a corrosive gas) and carbonate ion (which can form scale).

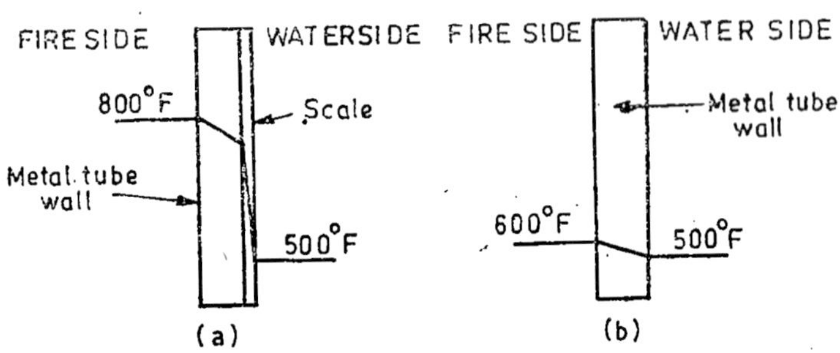

Fig. 5.3: Deposition acts to reduce heat transfer from combustion gases to boiler water. Deposits often provide enough insulation to cause tube metal overheating.

Sludgelike solids in boilers consist of various salts and oxides that precipitate in the bulk water and are transported through the boiler by normal water flow patterns. A minimal amount of sludgelike material can be controlled by normal boiler blowdown. One way to prevent crystalline scale formation in boilers is to chemically induce an amorphous, sludgelike precipitate.

Typically, boiler feedwater is a mixture of makeup and recycled steam condensate. Either or both of these water sources can contain iron and copper oxides, which can form particularly troublesome deposits in the steam-release areas of boilers.

Two major types of internal chemical treatment are precipitation with phosphate and solubilization with chelant. Polymeric dispersants are generally used to supplement both types of boiler water conditioning. In cases where water of particularly high purity is available, the polymers may be used alone.

An alternative form of treatment occasionally used in industrial boilers, but primarily designed forutility boilers of extremely high pressure and temperature, is an all-volatile treatment of ammonia and hydrazine. These operating conditions demand ultra-high-purity water, this type of treatment is preferable since it does not contribute any solids to the boiler water.

2. *Precipitation*

Phosphate is very stable at boiler water temperatures but extremely insoluble when it encounters calcium at high pH and high temperature. Calcium phosphate particles, in contrast to calcium carbonate, do not readily adhere to metal surfaces. For these reasons, phosphate is the standard internal treatment chemical.

In the phosphate form of treatment, alkalinity is maintained at a sufficient level to precipitate magnesium as magnesium hydroxide, which, in turn, adsorbs silica from the boiler water. The resulting sludgelike deposits, consisting of calcium phosphate and magnesium silicate, are further conditioned by polymeric dispersants to prevent their accumulation in the boiler.

For application of this treatment, a phosphate corrective such as disodium phosphate is added to the boiler water. If the natural alkalinity of the makeup water is insufficient, then an alkali such as caustic soda is added.

Polymeric dispersants are synthetic organic compounds of relatively low molecular weight and high charge density. Natural polymers such as tannins and lignins have been replaced by synthetic polymers and copolymers such as polymethacrylate, polymaleic anhydride, and sulphonated styrene maleic anhydride. The polymers function as dispersants by adsorbing on crystal nuclei, neutralizing surface charges, and preventing or controlling further crystal growth.

Blowdown of boiler water is adjusted to prevent undesirably high levels of dissolved solids and suspended solids that are circulated in the boiler water.

3. Chelation

Metal cations found in boiler water, such as calcium, magnesium, iron, and copper, can be solubilized to prevent deposition and scale formation. The functional agents in this type of treatment are known as chelants. Chelants are organic, anionic chemicals that form soluble compounds with the metals. The two chief chelating chemicals used in industrial boiler water treatment are ethylenediaminetetraacetic acid (EDTA) and nitrilotriacetic acid (NTA).

Chelants are added continuously in the stoichiometric quantity required to solubilize any metal contaminants present in the boiler water. The stabilities of the resultant metal—chelate reaction compounds vary considerably. EDTA has more coordination sites than NTA and forms stronger soluble complexes with the metal cations. However, EDTA in excess of that required for contaminant chelation can decompose in the boiler water and large excesses of EDTA and NTA can react with the magnetite (Fe_3O_4) film protecting the boiler. Other anions present in the boiler water (*e.g.*, phosphate, silicate, and hydroxide) tend to compete with chelants, limiting their effectiveness.

For these reasons, chelant treatment chemistry must be precise. Frequent analytical testing of the feedwater and internal boiler water is necessary to maintain a proper chelant feed rate.

Polymeric dispersants are used to supplement chelant treatment. Polymers are particularly effective in dispersing metal oxides and sludges that form due to chelant instabilities. Proper application of a chelant—dispersant treatment results in clean heat transfer surfaces and efficient boiler operations.

C. Blowdown Control

Guidelines for boiler water quality have been established through years of operating experience with boilers of varying design. Blowdown frequency is determined on the basis of maintaining boiler water within these guidelines to prevent deposition and to provide good-quality steam. Total solids, alkalinity, and silica as well as the iron, copper, or total hardness content of the feedwater affect the blowdown rate. A partial list of guidelines published by the American Society of Mechanical Engineers (ASME) is given in Table 5.5.

TABLE 5.5

Partial List of Guidelines for Boilers

Drum pressure (psig)	*Silica (ppm SiO_2)*	*Total alkalinity (ppm $CaCO_3$)*	*Specific conductance (μmho/cm)*
0-300	150	700	7000
301-450	90	600	6000
451-600	40	500	5000
601-750	30	400	4000
751-900	20	300	3000
901-1000	8	200	2000
1001-1500	2	—	150
1501-2000	1	—	100

Two types of blowdown, manual and continuous, are practiced in the operation of most industrial boilers. Manual blowdown consists of repeated, short openings of a valve located at a low point in the boiler (usually the mud drum) to blow boiler water with entrained solids from the unit. Performed intermittently, it is primary designed to remove sludge formed in the boiler water; it also aids in controlling dissolved solid concentrations.

The need for manual blowdown is minimized or eliminated completely with high-quality feed-water and a solubilizing chelant treatment. Specific directions for manual blowdown are provided by the boiler manufacturer.

Invariably, a system of continuous blowdown is applied to modern industrial boilers. As the name implies, water is removed from the system at a low, steady rate to maintain a uniform boiler water chemistry. A continuous blowdown line is installed at a point several inches below the normal water level in the steam-release drum. This is a point of maximum concentration of boiler water solids.

Chemical analysis of boiler water is performed on a sample from the continuous blow-down line. Controls over the treatment programme usually consist of analytical determinations for:

1. oxygen scavenger residual,
2. phosphate or chelant,
3. alkalinity or pH,
4. silica, and
5. blowdown.

Specific conductance of the boiler water is a common method of control over blowdown. Chloride is also widely used if the makeup water is not demineralized.

Total btowdown, the combination of manual and continuous blowdown, is dependent on water chemistry, type of boiler, steam load characteristics, and many other operating conditions.

Heat recovery of boiler blowdown streams is an important consideration in the economics of operation of a boiler system. A significant amount of high-temperature water is involved in the average industrial system. With continuous blowdown facilities, heat recovery can be readily accomplished. Auxiliary equipment, such as illustrated in Fig. 5.4, is used to produce steam at lower pressure for recycle and also exchange the heat from the residual blowdown water.

D. Carryover

1. Causes and Effects

Steam purity and steam quality have different, but related, meanings. Steam quality is simply the amount of moisture in the steam. Steam purity refers to the amount of solid, liquid, or gaseous contamination in the steam. Steam purity is a major operating parameter in industrial boiler systems. Units designed with superheaters, systems with steam turbines, and processes that need steam all demand that the carryover of boiler water solids with steam be kept to a minimum.

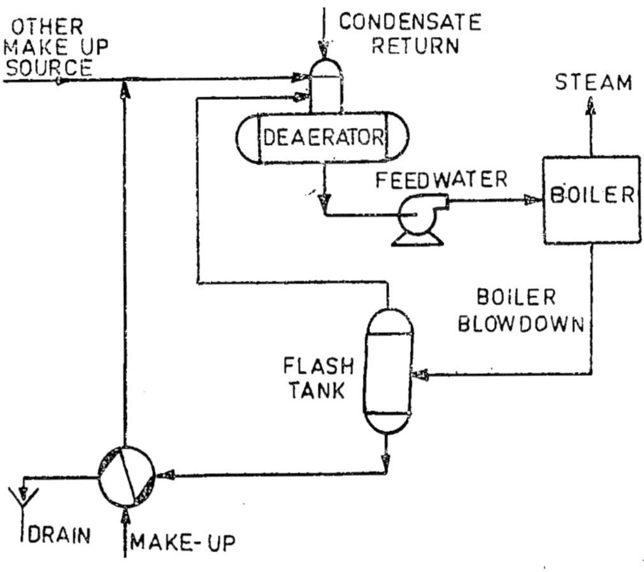

Fig. 5.4: Typical boiler blowdown heat recovery system using a flash tank and heat exchanger

Carryover is any contaminant that leaves the boiler steam drum with the steam. The most common carryover is entrained boiler water droplets that retain dissolved and suspended solids. However, carryover can also include gases and solids that vaporize with the steam. For example, silica will vaporive significantly when boiler pressures exceed 600 lb/in.^{2}

Where superheaters and turbines are involved, steam purity of less than 30 ppb of total solids may be necessary to prevent harmful deposits. Solids that carryover with the steam can form deposits in superheaters, leading to the overheating and rupture of tubes. Further down-stream, deposits can decrease the efficiency of operation of turbines and condensers. Silica carryover can result in tenacious deposits on turbine blades, jeopardizing the operation of this type of equipment.

Carryover can be caused by mechanical or chemical factors. Boilers are usually designed with internal steam-purifying devices of some type that are, in some cases, supplemented by external steam-purifying units. These types of devices can vary in sophistication to provide practically any degree of steam quality. However, most boiler manufacturers' guarantee of 1 ppm of total solids can be surpassed with attention to proper boiler water chemistry. During operation these units can become defective, leading to one of the chief causes of mechanical carryover. Other mechanical factors to consider are boiler water loads, water level, method of firing, and boiler design.

The chief chemical causes of carryover are foaming and vaporization of solids. Foaming is the formation of bubbles on the surface of the boiler water. Where the foam is stabilized, it can completely fill the steam drum and contribute carryover of solids to the steam when the bubbles burst. Foam can be caused by high dissolved solids, high alkalinity, and suspended solids. One of the chief stabilizers of objectionable foam is the suspended solids in the boiler water. Synthetic detergents, oil, and other contaminants present in boiler makeup water can contribute to foaming.

2. *Control Methods*

Gravity separation is sufficient for production of good-quality steam at lower boiler operating pressures (*e.g.*, 150 lb/in.^{2}), where the steam and water density differential is great and the steam drum is adequate for the steam load.

Antifoam agents can significantly reduce carryover caused by chemical factors. Compounds classified as polyalcohols and polyamides are particularly effective in minimizing entrainment of boiler water with steam.

Limiting boiler water solids through blowdown control is the principle method of minimizing carryover. However, oil and other organic contaminants can cause foaming even with minimal levels of boiler water solids. In many cases, contamination must be eliminated. Chemical antifoam agents can sometimes compensate for minor organic contamination and can permit lower blowdown rates than called for by the guidelines.

3. Steam Purity Measurement

Accurate methods of steam sampling and analysis are available for measuring impurities. The more advanced techniques permit analysis in the parts per billion range, which is necessary for some high-pressure industrial operations.

The most common method of determining steam purity is by means of specific conductance. The conductivity of water, measured in micromhos (μmhos), is an indication of its ionic content. Gases present in the condensate, such as ammonia and carbon dioxide, ionize in water and can interfere with the measurement of steam purity. The gases can be eliminated by degassing or by ion exchange. Specific conductance is not practical when steam impurities of 1 ppm or less are involved.

Sodium typically constitutes one-third of the total solids in boiler water and is the highest single constituent. Extremely low concentrations of sodium can be readily detected by specific ion analysis and flame spectrophotometry of condensed steam samples.

Steam sampling can be accomplished in a representative manner with the use of sampling nozzles as specified in ASTM (American Society of Testing Materials) Standard D 1066. Saturated steam is usually sampled for steam purity studies, although superheated steam can be sampled with the proper sample point desuperheating arrangement.

E. Condensate Systems

1. Causes of Corrosion

The condensate system is subject to corrosion from oxygen,

carbon dioxide, and ammonia. Condensate return lines are constructed primarily of mild steel and can be extensive in total length throughout the plant. Condensing equipment in the steam system usually includes copper or copper alloys because of its favourable heat transfer characteristics. Numerous other metals can also be present in the system. The metal oxide corrosion products carried back to the boiler impose an added burden on the boiler system. Any process chemicals or cooling water that might contaminate the condensate system through corrosion-induced leaks impair the reliability of the boiler system. For these reasons, condensate return line systems require proper maintenance.

Oxygen can enter the condensate system via incomplete deaeration of the boiler feedwater or through leakage of seals in condensate pumps. Pressure changes from intermittent operation of boiler systems also tend to pull oxygen into the pipes. Typically, oxygen attack is characterized by localized pitting of metal, which results in rapid loss of equipment.

Carbon dioxide forms an acid in condensate systems that dissolves metal piping in a generalized manner. Failure usually occurs first at threaded sections of piping, where the metal is thinner.

Carbon dioxide can enter the condensate system from either incomplete degasification of the feedwater or the breakdown of bicarbonate and carbonate alkalinity. The heat of the boiler produces 0.79 ppm of carbon dioxide for each ppm of sodium bicarbonate in the feedwater and 0.35 ppm of carbon dioxide from each 1 ppm of sodium carbonate (as $CaCO_3$) in the feedwater.

Ammonia is sometimes fed directly to high-pressure steam sytems to elevate the condensate pH. Careful control is required to prevent overfeed that could result in corrosion of copper metallurgy. Ammonia can also result from the breakdown of nitrogenous organic contaminants and partial decomposition of hydrazine.

2. *Chemical Treatment*

Two classes of chemical agents are used to protect condensate systems from corrosive attack: neutralizing amines for carbon dioxide control and filming amines for oxygen and carbon dioxide control.

Neutralizing amines are the most widely used chemicals for the protection of condensate system. The amines hydrolyze in the condensate, contributing hydroxyl ions, which neutralize the carbonic acid formed from CO_2. Continuous feed of neutralizing amine is controlled to elevate the condensate pH to the range of 8.5 to 9.0. Commonly used neutralizing amines include morpholine, cyclohexylamine, diethylaminoethanol, and dimethylisopropanolamine.

One of the original methods for protecting condensate piping from carbon dioxide and trace oxygen attack was the use of a filming amine such as octadecylamine. This type of fatty acid amine forms a film on the metal surface that acts as a barrier to attack from both oxygen and carbon dioxide. However, high levels of oxygen tend to disrupt the film and cause fouling. Filming characteristics of this longchain amine are often improved by adding an emulsifying agent or a neutralizing amine to the product.

Octadecylamine is usually fed at the rate of 0.5 to 1.5 ppm based on the rate of steam production. Filming amines are fed directly to the steam distribution system rather than to the boiler. Overfeed of filming aminies and contamination from multivalent ions can cause deposits in the return line system.

Oxygen scavenging agents can be fed directly to the condensate where air leaks into the system. Reducing agents such as hydroquinone react with iron oxides to form a protective magnetite layer for steel piping.

IV. COOLING WATER SYSTEMS

A. Types of Systems

Cooling water systems are used throughout industry for the removal of heat from all types of processes and equipment. They are all designed with heat exchangers of various types and metallurgy and with piping to transport the cooling water. The function of the cooling water is to absorb heat at a desired point in the system and transport it to a location where it can be dissipated or exchanged.

There are three distinctly different types of cooling systems: open recirculating, closed, and once-through. Each of these systems requires a different approach as far as a treatment programme is concerend, since potential problems vary in intensity from one type of system to the other. These water treatment programmes are designed to prevent scaling, corrosion, and fouling problems that might impair the efficiency of operation or cause shutdown or premature failure of any portion of the system.

1. Open Recirculating Systems

In industry, the most widely used cooling water system reuses water after heat is dissipated in a cooling tower, spray pond, or evaporative condenser. Since water is continually reused, this type of system, in contrast to a once-through mode of operation, reduces the amount of makeup water needed, resulting in significant water conservation.

Use of the high latent heat of vaporization of water is encouraged in the design of open cooling systems. The sensible heat exchange in a cooling system can be significant in colder temperatures but is usually only a small percentage of the overall heat rejection from the cooling water. Evaporation is the principal mechanism for heat rejection. Approximately 1000 Btu is lost from the water for every pound evaporated. This is equivalent to a 1 per cent evaporation loss of water for each 10°F temperature drop with respect to the quantity of water circulated.

When pure water is evaporated from the cooling system, dissolved solids in the circulating water are concentrated. The term cycles of concentration is used to indicate the degree of concentration of the circulation water solids as compared with the makeup water. At 2.0 cycles of concentration, for example, the circulating water has twice the number of solids by weight as the makeup water. Since it is usually present in makeup supplies, chloride ion is frequently used as a tracer for determining cycles of concentration. For example, if the chloride in the circulating water is 100 ppm and the makeup water contains 50 ppm of chloride, then cycles of concentration are 2.0.

Cycles of concentration are an important parameter in the operation of open cooling systems. Cycles of concentration that, are developed too high can result in scaling and corrosion, and cycles that are low can have an adverse impact on the economics of the operation.

Cooling towers are most commonly used in open systems for dissipation of heat. Spray ponds are less efficient and are rarely used. In some open cooling systems, a lake or pond is used as a heat sink, and some evaporation of the recirculation water is obtained from natural climatic conditions. Numerous evaporative coolers or condensers are also used in which cooling water is circulated directly over process piping to evaporate some of the cooling water.

Cooling towers are constructed of various materials, including wood, metal, and concrete. They are designed to effect the intimate contact of water and air for maximum evaporation. Cooling water towers are described by the type and pattern of air flow. Natural draft towers rely on a "chimney effect" to obtain circulation of the air through the water. Mechanical draft towers use fans to move the air and are subdivided into forced-draft or induced-draft towers. The tower is an induced design, since the fan is located on top of the tower and draws the air through it. In forced-draft cooling towers, the air fan is located at the bottom of the tower.

Various types of packing or fill are used in cooling towers and are classified as either splash or film types. A typical efficient cooling tower can provide a 10°F approach to the wet bulb temperature. The wet bulb temperature of dew point of the air is the lowest temperature to which water can be cooled by cooling tower water evaporation. However, it is not economically practical to design a cooling tower to lower the cooling water temperature equal to the wet bulb. The difference between the sump temperature and the wet bulb is referred to as the approach.

In the operation of modern cooling towers some amount of water is entrained with the air, and this loss of the liquid water from the tower system is known as windage. In mechanical draft towers, the windage loss is usually in the range of 0.05 per cent to 0.3 per cent of the amount of water circulated over the tower.

The water level in a cooling tower system is maintained constant by the flow of makeup water, which compensates for the loss of water from evaporation, windage, and blowdown. Blowdown is the removal of a stream of concentrated circulating water from the tower system to limit cycles of concentration. The formula defining cycles of concentration is

$$\text{cycles of concentration} = (\%E + \%B)/\text{per cent } B$$

where per cent E is the evaporation loss expressed as a per cent of the circulation rate and per cent B the combined windage and blowdown loss expressed as a per cent of the circulation rate.

Since scale, corrosion, and fouling are intensified in open recirculation systems as a result of their exposure to the atmosphere and reuse of cooling water, treatment programmes are very important for trouble-free operation. For good economics, it is important to minimize any water losses from the system over and above the blowdown and windage designed for the programme.

Proper maintenance of the cooling water system equipment is also necessary. Since cooling towers are usually constructed of redwood, douglas fir, and other species that are subject to deterioration, a proper maintenance programme should include inspection and replacement of the tower wood.

2. *Closed Systems*

These systems reuse water circulated in a cycle of alternate cooling and heating with little or no evaporation or makeup water. Ideally, zero makeup water is required for these systems. If makeup water exceeds 1 per cent, then the system should be considered as a once-through system.

In the closed cooling system, heat transferred from the process or equipment is rejected in a water-to-water or water-to-air exchanger. An example of a closed system is the automobile cooling system, in which heat from the engine is dissipated to the atmosphere through the radiator.

Closed systems are often used for critical applications where complete reliability is required. The chilled water circuits of air-conditioning units and compressor cooling systems are examples of systems that are commonly closed with respect to makeup water.

Makeup water for closed systems is generally of high quality, with steam condensate, zeolite-softened water, and city water being preferred in that order. Scaling is not usually a problem in closed systems because of the closed nature of the system. Corrosion is a potential problem, however, and a good inhibitor treatment programme is necessary. Proper design of the system should eliminate the possibility of galvanic corrosion by the use of insulators where dissimilar metals are involved.

3. Once-Through Systems

Once-through systems use water from a particular source on a one-time basis without recirculating the water for reuse. Some once-through systems use a domestic supply as a source of makeup water. More commonly, plants located near a water supply, such as a lake, river, or sea, use these sources for once-through cooling. Occasionally, a well-water supply that is at a low, constant temperature, such as 60°F, is used on a once-through basis for cooling processes or products. Occasionally, a well-water supply used as makeup to a cooling tower system is utilized on a once-through basis for some heat-exchange service. In these cases, the treatment guidelines for both kinds of systems must be considered.

Once-through cooling water systems are quite common, but many have been replaced with recirculating systems. Even where there is a plentiful surface water supply, a once-through system can impose the danger of thermal or chemical pollution on the receiving stream. Makeup water is elevated in temperature to different degrees in various once-through cooling systems, but there is no significant evaporation involved. The problem of corrosion is generally handled in the design stage by specifying alloyed metallurgy resistant to attack. Control over scale and biofouling are usually the main points involved in water treatment programmes for once-through systems. [See *Wastewater Treatment.*]

B. Types of Problems

1. Corrosion

This discussion on corrosion will focus primarily on mild steel, which is present in most cooling systems. The corrosion mechanisms described, however, can apply to many different types of metallurgy. Some reference will also be made to copper alloys and stainless steels, since these too are used in may industrial cooling water systems.

Corrosion is an electrochemical process in which metal loss occurs at an anode and the resulting electrons produced are used at the cathode. The cooling water serves as the electrolyte to complete the electrical circuit. The amodic reaction is

$$Fe^0 = Fe^{2+} + 2e^-$$

Several reactions can take place at the cathode, depending on pH and other conditions. Primarily, oxrgen is reduced at the cathode according to the reaction

$$O_2 + 4e^- + 2H_2O = 4OH^-$$

Iron, which comes into solution at the anode, reacts with the hydroxyl ions and precipitates on the metal surfaces. The iron oxide formed is approximately nine times as voluminous as the base metal. These formations, called tubercles, can have the secondary effect of impeding cooling water flow.

Corrosion can occur in a uniform manner or can be localized in nature. The formation of anodic and cathodic sites of corrosion can be initiated by differences in surface structure or composition. In cooling systems, corrosion is usually of a localized nature, such as pitting attack. Other forms of localized corrosion are galvanic, crevice, intergranular, erosion, and stress corrosion cracking.

Pitting, which can be initiated by oxygen and other corrodents in the water, is a serious from of corrosion that can cause rapid failure of heat-exchanger tubing. Once a pit starts, the action is self-sustaining because of the hydrogen and chloride ions that concentrate within the pit.

Galvanic corrosion is caused by the contact of two dissimilar metals. The driving force for the corrosion is the potential difference that develops between the two metals. The galvanic series of metals is given below.

Galvanic Series of Metals (partial list)

Corroded end	Lead
(anodic or least	Tin
noble)	Nickel (active)
Magnesium	Brasses
Zinc	Copper
Aluminium 2S	Bronzes
Cadmium	Copper-nickel alloys
Steel or iron	Titanium
Cast Iron	Monel
Chromium-iron	Silver
(active)	Graphite
18-8-Cr-Ni-Fe	Protected end
(active)	(cathodic or most noble)

Corrosion can occur at a crevice or similar formation in a cooling system. In this case, the solution in the crevice differs in oxygen concentration from the flowing bulk water and so creates an anodic site. Attack takes place in a manner similar to that of a pit. Crevices should be eliminated from a cooling system wherever possible.

Underdeposit corrosion results from the formation of concentration cells beneath deposition in a cooling system. This type of corrosion is prevalent in improperly treated systems. Elimination of deposits is a course of action for prevention of this type of attack.

Intergranular corrosion in cooling systems results from attack on metal alloys by corrodents in the circulating water. This is a form of localized attack that occurs at the metal grain boundaries and most commonly involves stainless steels or copper alloys.

Erosion/corrosion is caused by high water velocity and suspended solids in circulating cooling water. It involves an increased rate of corrosion due to mechanical effects.

Stress corrosion can occur in certain metals due to a combination of tensile stress and a corrosive environment. Stress corrosion cracking is not usually found in cooling water systems because temperatures in excess of 180°F are usually required. Cooling water systems are generally designed with an average maximum bulk water temperature of 120°F. Metals used in cooling systems that are susceptible to stress corrosion cracking are austenitic stainless steels and brasses. The presence of chloride and oxygen is necessary to initiate stress cracking of susceptible metals.

The concentration of dissolved solids, especially chlorides and sulphates, in circulating cooling water is a factor in potential corrosion. Corrosion increases with increased conductivity of the cooling water. The effect of temperature is to increase corrosion at a given oxygen concentration as the temperature increases.

The presence of corrosive gases, such as hydrogen sulphide, sulphur dioxide, and ammonia, can also be a factor in corrosion. Ammonia is particularly corrosive to copper and copper alloys.

Corrosion occurs at a rapid rate at low pH due to the presence of hydrogen ions. When the pH is below 6.0, the water is generally very corrosive. At pH 8 and above, the corrosion rate of steel is substantially lower. The use of chlorine, which tends to lower pH, can result in higher corrosion rates, particularly when the circulating water is low in alkalinity. Where sulphuric acid is used for pH control of the system, any overfeed can result in severe corrosion.

Since heat exchangers are integral parts of processes and are relatively expensive, corrosion that significantly shortens their service life is usually not acceptable. Heat exchangers of all types of construction are found in cooling systems, including plate and frame, shell and tube, spiral, jackets, and pipe exchangers. The complex geometry of heat exchangers is an added incentive for good corrosion control.

2. Scale

Scale is the precipitation of salts from solution with the formation of crystalline adherent deposits on cooling system surfaces. The solubilities of most scale-forming salts in water decrease with increasing temperature. Frequently, the rise in water temperature due to cooling service is sufficient to cause precipitation of scaling material even in once-through systems. Calcium carbonate, calcium sulphate, magnesium silicate, iron, and manganese are the most common scales found in cooling systems Temperature, pH, and concentrations of solids are the principal factors involved.

The Langelier index is used as a qualitative measure of the potential for calcium carbonate scale formation in cooling systems. The pH of saturation of calcium carbonate is determined by use of this index, which is based on the factors of pH, calcium, total alkalinity, dissolved solids, and temperature. The algebraic difference between the actual pH of water and its calculated pH of saturation is the saturation index, given by

$$\text{saturation index} = \text{pH} - \text{pH}_s$$

Qualitatively, a positive index indicates possible calcium carbonate deposition, and a negative index indicates a corrosive tendency with no calcium carbonate scaling.

Subsequent to the work of Langelier, a stability index was developed by Ryznar to provide a more quantitative measurement of scaling:

$$\text{Stability Index} = 2(\text{pH}_s) - \text{pH}_a$$

A stability index of 6 or less indicates a decreasing tendency for corrosion and a scaling increase, whereas an index of 7 or above indicates decreasing scaling tendencies and a corrosion increase. Use is made of these primarily to compare the tendencies of one water with another.

Iron scales can develop in cooling systems and can be extremely troublesome. Many well waters contain iron in a soluble ferrous bicarbonate form. Upon exposure to oxygen in the cooling system, the ferrous iron is oxidized to the insoluble ferric state, Iron and manganese scales can also originate from surface supplies where the metals are held in solution by the activity of natural organic substances. Upon chlorination in the cooling system, the iron and manganese complexes are destroyed, with resultant precipitation of scale.

Calcium sulphate and silica have limited solubility in cooling water and are potential scale formers. Through concentration in open recirculating systems, the solubilities can be exceeded. Calcium sulphate has inverse solubility and will precipitate initially at the highest heat-transfer surface. Silica is more soluble at higher pH and temperature, but the concentration is usually limited to 150 ppm in circulating cooling water.

3. Fouling: Biological and Chemical

Fouling of cooling systems is usually defined as deposition of non-scale-forming substances, such as:

1. mud, silt, and suspended matter in makeup water,
2. chemicals from process contamination,
3. particulate matter and insects from the atmosphere,
4. migrating corrosion products, and
5. products of biological activity.

These potential sources of deposition are differentiated from scale so that a more direct approach can be taken to control a specific problem. However, scaling and general fouling are similar in that they can seriously interfere with heat transfer and block water circulation.

Biological fouling, often referred to as slime, is caused by the growth of algae, fungi, and bacteria. Slime is an accumulation of microorganisms and their metabolic products. The biomass that is formed can trap organic and inorganic debris. Temperature, pH, and nutrient content of cooling water are usually ideal for the rapid development of many forms of microbiological life. Organisms can be constantly injected into the system from the makeup water, from process contamination, and from the atmosphere. In open cooling tower systems, sunlight is also available for development of algal growths. Totally, large and varied microbial populations can develop in cooling systems without treatment.

In brackish water and seawater systems, marine organisms are a source of fouling. In some fresh-water systems, aquatic plants and aquatic animals, such as clams, can interfere with operation.

Algae use photosynthesis for development and can form large masses on open areas of cooling systems. Algae can interfere with water circulation on cooling towers and slough off to cause problems in heat exchangers. Fungi can contribute to the biomass and attack cooling tower wood.

Several types of bacteria are particularly troublesome in cooling systems. Iron bacteria, sulphate-reducing bacteria, nitrifying bacteria, and spore-forming bacteria are only part of the possible microflora complexes that can develop.

Iron bacteria contaminate well water and surface supplies. Sulphate-reducing bacteria are anaerobes and can exist under deposits in cooling systems. Sulphate reducers form hydrogen sulphide, which is very corrosive to most metallurgy.

Nitrifying bacteria, soil organisms that can enter cooling systems, oxidize ammonia to nitric acid. Low pH and corrosion can develop from this type of activity.

Spore-forming bacteria can survive severe physical changes in the environment and even toxic conditions. When conditions are right for their development, such as in a typical cooling system, the spore formers can multiply rapidly to form a biomass.

Chemical slimes are often difficult to distinguish from slimes of microbiological origin. Various metals, such as aluminium and iron, form insoluble, gel-like salts in cooling water. Silts and dust particles from the atmosphere enter a cooling water in colloidal form. Although these small particles carry electrostatic charges and tend to repel each other they can grow under the thermal and chemical conditions of the cooling system and result in voluminous deposits.

Open recirculation systems tend to have more operating difficulties from lack of good fouling control than from other potentially troublesome areas. Usually, a corrosion inhibitor and scale-control programme are maintained properly, but heat transfer becomes dificient through chemical or microbiological fouling. These types of fouling can occur rapidly, and critical equipment can fail to obtain the proper cooling.

C. Treatment

1. Corrosion Control

Corrosion of common metallurgy in cooling systems is primarily controlled by the use of chemical inhibitors. The two types of inhibitors most commonly employed are passivators and barrier formers. Inhibitors can also be classified as anodic cathodic, or a combination of both. Anodic inhibitors control the reaction at the anode, while cathodic inhibitors control the reaction at the cathode. Both of these reactions were discussed in Section I.

Inhibitors that form a barrier to effect corrosion control are usually cathodic. At the cathodic site of a corrosion cell, the pH is elevated by the formation of hydroxyl ion. Compounds, such as calcium carbonate or calcium phosphate, are then selectively precipitated at the cathode to prevent further cell activity.

Industrial corrosion inhibitors are primarily inorganic compounds and include chromate, phosphate, nitrate, molybdate, silicate, and zinc. Some organic phosphorus compounds, such as hydroxyethyldenedisphosphonate (HEDP), function as corrosion inhibitors for steel, particularly at higher pH. Other organic inhibitors include tolyltriazole and benzotriazole, which are very effective for protection of copper and copper alloys.

Chromate is an extremely efficient inhibitor, but its usage has been de-emphasized in recent years because of its toxicity. Chromate can be used at high levels or combined with other inhibitors and dispersants to minimize treatment concentration. The criteria for a good industrial inhibitor or inhibitor combination is to limit corrosion of steel to less than 2 mils/yr, which is equivalent to 0.002 in./yr. This figure is based on a uniform corrosion rate, and pitting should be nonexistent. Combinations of chromate-base treatments and phosphate-base treatments are used widely within these criteria.

The combination of chromate (a good anodic inhibitor) with zinc (a good cathodic inhibitor) has been used widely with pH limited to a range of 6.0 to 7.0. With many makeup water supplies, the pH of 6.0 to 7.0 requires acid to be fed to eliminate some of the natural alkalinity of the makeup water. This type of programme is listed as number 2 in Table 5.6.

Zinc functions as a cathodic inhibitor by combining with hydroxyl ion produced by the corrosion cell. The zinc hydroxide is insoluble, precipitates, and stifles further cell activity. When phosphate is present, such as treatment number 3 in Table 5.6, zinc phosphate is the cathodic inhibitor. Zinc phosphate is less soluble than zinc hydroxide and is a more efficient inhibitor. In corrosive waters, pitting can often be eliminated by utilizing zinc phosphate with chromate.

The chromate-phosphonate combination in Table 5.6 is an example of high-pH chromate treatment where phosphonate is used to obtain cathodic inhibition. In this system of treatment, calcium carbonate, naturally present in the water, is a cathodic inhibitor. Calcium carbonate activity is controlled by the phosphonate.

The current trend is to elevate the pH of the circulating water in open cooling systems by eliminating or decreasing the feed of acid. With the use of some recently developed copolymers, the zinc can be maintained in solution at elevated pH. This combination, listed as number 5, in Table 5,6 gives excellent corrosion protection.

TABLE 5.6

Chromate-Based Treatments

Number	*Inhibitor*	*Treatment corrosion*	*pH*
1.	Chromate	200-500 ppm as CrO_4	6.0-8.0
2.	Chromate-zinc (Zn)	10-30 ppm CrO_4, 2-5 ppm *Zn*	6.0-7.0
3.	Chromate-zinc phosphate	10-30 ppm CrO_4, 2-5 ppm Zn, 2-5 ppm PO_4	6.0-7.0
4.	Chromate-phosphonate	5-10 ppm CrO_4, 2-5 ppm phosphonate	7.5-8.5
5.	Chromate-zinc copolymer	10-20 ppm CrO_4, 1-2 ppm Zn, 5-10 ppm copolymer	7.0-8.5

Polyphosphates, which are molecularly dehydrated orthophosphates and include pyro-and metaphosphates, have been used in water treatment for many years. In water solution, the Polyphosphates are hydrolyzed back to the natural orthophosphate state. Polyphosphate is a mixed inhibitor in that it is cathodic when combined with calcium and anodic in the orthophosphate form. Orthophosphate is an extremely effective anodic inhibitor, rapidly promoting the formation of protective gamma iron oxide on iron surfaces in cooling systems. Widespread industrial use has been made of the orthophosphate-base corrosion inhibitors since the advent of a copolymer that can prevent calcium phosphate scale. A typical system of treatment is as follows: ortho- and polyphosphate, 10-20 ppm; copolymer, 5-10 ppm; pH7.0-8.0.

With the phosphate-base treatment, efficient heat exchange is obtained because of the adherent oxide film. The role of the cathodic inhibitor, calcium phosphate, is minimized to provide a clean system by balancing the phosphate, pH, calcium, and copolymer concentrations.

Nitrite and nitrite-based inhibitors are used in both open and closed cooling systems. Nitrite is an anodic inhibitor, but very high concentrations are necessary (500-1000 ppm) to obtain good protection. Therefore, nitrite-based treatments are primarily limited to closed systems in which there is no water loss from the systems. Since oxidizing agents and certain bacteria will attack nitrite, proper precautions and testing are necessary to prevent loss of the inhibitor.

Silicate and polysilicates are weak anodic inhibitors that are occasionally used for corrosion protection where chromate, phosphate, and zinc cannot be employed for environmental reasons.

With the trend to operate open systems under alkaline conditions, the pH is usually in the range of 7.5 to 9.0. At this high range, the corrosivity of the water is decreased, compared to low-pH systems. Calcium carbonate crystal inhibitors also function in this pH range to give some cathodic corrosion control without scaling. Some modern alkaline treatments, such as combinations of zinc, phosphonate, and copolymer, are highly effective corrosion inhibitors. Acid feed is eliminated with the alkaline treatment, which is a distinct advantage.

In once-through industrial cooling systems and domestic water systems, polyphosphates are used for some added corrosion protection. Polyphosphate effectively reduces tuberculation in piping and can minimize red water caused by corrosion products of iron. Effective results have been obtained with treatment levels as low as 2 ppm. Corrosion protection at this low treatment level can be improved by adding a small amount of zinc (1 ppm or less).

For positive protection of closed cooling systems, high levels of chromate treatment or nitrite-base treatments are employed. Buffered chromates in the range of 500 to 1000 ppm of chromate and usually satisfactory.

Nitrite-borate treatments and nitrite-molybdate are widely used, particularly where chromate is not acceptable. Nitrite concentrations in the range of 500 to 1000 ppm as NO_2 are desirable for protecting steel metallurgy. Combinations of nitrite-molybdate are advantageous in that nitrite is a rapid passivator of steel. Molybdate can maintain protection of the system if nitrite is lost through chemical or biological activity. Molybdate is usually maintained in the range of 200 to 250 ppm as MoO_4. Molybdate, which is in the same chemical family as chromate, is an anodic inhibitor that is not as efficient as chromate. However, it is colourless and nontoxic, and when used with other inhibitors (such as nitrite), it serves a good purpose.

2. *Scale Control*

Chemical treatments used for scale control in cooling systems are acids and crystallization inhibitors. Blowdown is also used in conjunction with the chemical treatments to eliminate scaling conditions in open recirculating systems.

Sulphuric acid is normally employed to control alkalinity in cooling water. Other acids, such as hydrochloric or nitric acid, can be used but are usually more expensive and present other difficulties. Carbon dioxide, which forms carbonic, acid in solution can be used to control the pH of an open recirculating system. However, it is rarely used because it readily vaporizes from the water in passage over a tower.

Sulphuric acid reacts with the natural bicarbonate alkalinity of the makeup water according to the reaction

$$Ca(HCO_3)_2 + H_2SO_4 \rightarrow CaSO_4 + 2CO_2 + 2H_2O$$

The carbon dioxide generated in this reaction is released from the water so that there is no increase in the dissolved solids content of the acid-treated water. The calcium sulphate that is formed has greater solubility than calcium carbonate.

Crystallization inhibitors used for scale control include polyphosphates, phosphonates, phosphoric acid esters, and low-molecular-weight polymers.

In domestic water systems and in once-through cooling systems, polyphosphates have been traditionally used as 2.0 ppm for controlling calcium carbonate scaling. Polyphosphate permits the development of a slightly positive saturation index in the treated water without the danger of scale formation.

In open recirculating systems, phosphonates and low-molecular-weight polymers are used extensively for scale control. The phosphonates, including hydroxyethylidenendiphosphonate and aminotrimethylenephosphonate, have found wide acceptance for controlling calcium carbonate scale. These agents are more stable than the polyphosphates and permit the development of a higher degree of supersaturation with respect to calcium carbonate.

Operating limits for potential scale-forming substances in open cooling systems vary considerably with the numerous types of industrial processes requiring cooling water survice. However, some generalization of industrial operating limits would be as follows: calcium as $CaCO_3$, 1000 ppm; calcium sulphate ($[Ca^{2+}] \times [SO_4^{2-}]$), 1,750,000 ppm; magnesium silicate ($[Mg^{2+}] \times [SiO_2]$), 35,000 ppm; silica, 150 ppm.

The engineering approach to preventing scale in open cooling systems is to use the above limits with the appropriate scale inhibitor and the proper blowdown rate. The blowdown of water from the system would be at a continuous rate so as to maximize cycles of concentration without exceeding the limits given.

3. *Fouling Control*

Chemical fouling in open cooling systems is generally controlled by the use of dispersants and surfactants. The crystal growth inhibitors previously described also act as dispersants. Phosphonates, polymers, and polyphosphates are the principal fouling-control agents.

Surfactants or wetting agents are used to assist the dispersant activity where cooling water is contaminated by process chemicals, such as oil and grease. The wetting agents are also effective in assisting the dispersion of products of biological activity. Surfactants are usually nonionic polyalcohols, or polyethers and a typical feed rate is 10-30 ppm.

Microbiological growth can be controlled by chlorination and the application of cooling water biocides. Most microbiological life forms can be killed by chlorine treatment if exposed to a sufficient residual for a long enough time. The mode of action of chlorine involves oxidation of the cell. A free residual of 0.2 to 0.8 ppm is sufficient. This can be applied intermittently or continuously, depending on conditions in the open recirculating system. A typical chlorination programme would be applied 3 days per week for 4 hr on each occasion of treatment.

In once-through systems, chlorine is applied at the same residual, typically for 30 min to 2 hr daily. However, in view of the high toxicity of chlorine and chlorinated hydrocarbons, discharge of chlorine-treated water into public streams and water courses is closely regulated. Hypochlorites, chlorine dioxide, and chlorine donors are also used for slime control. Other oxidizing compounds are ozone and bromine-base products. These latter chemicals are only used to a minor extent in industrial cooling systems.

Many generic and proprietary biocides are used as alternatives or supplements to chlorination. Most of these have a different mode of activity from the oxidizing nature of chlorine. Proprietary biocides are often based on the following compounds: quaternary ammonium salts, methylenebis(thiocyanate), dibromonitriloproprionamide, isothiazoline, bromonitrostyrene, and bis(sulphone).

Proprietary biocides usually contain more than one compound so that a broad effect on controlling bacteria, fungi, and algae is obtained. On the other hand, some biocides are very specific, having, for example, more effect as an algicide than a fungicide.

In open recirculating systems, proprietary biocides are often used as a supplement to chlorination. Occasionally, the biocides are used alone to maintain control over biofouling. Usually, the proprietary biocide is added in a shot fashion, such as 10 to 30 ppm, to the cooling water. The approach is not to obtain complete sterilization of the system, but to control microbiological populations at a tolerable level. The biocide concentration is effective for a period of time depending on the blowdown rate. The theoretical depletion of a slime-control agent from the system can be determined from the formula

$$\log C_f = \log C_i - (BD\text{-}T/2.303\ V)$$

where C_f is the final concentration, in parts per million; C_i the initial concentration, in parts per million; BD the blowdown and windage loss, in gallons per minute; V the system capacity, in gallons; and T the time, in minutes. A programme that has been found practical is to repeat the shock treatment when C_f is 25 per cent of C_i. On this basis, the formula is simplified to

$$T = 1.385V/BD$$

This time is often referred to as the retention time in a cooling system, although it reflects only a 75 per cent depletion of a substance that has been shot fed to the system. Some nonoxidizing biocides are effective over this period of time.

Still other industrial biocidal compounds are advantageous in that they spontaneously break down into nontoxic fragments a short time after being added to the water. This characteristic is advantageous in that after the biocidal activity is accomplished, the compound is no longer toxic. Among the biocides that have this activity are methylenebis(thiocyanate), dibromonitrilopropriοnamide and bromonitrostyrene.

D. Monitoring

Monitoring a cooling system for scale, corrosion, and fouling control is practiced regularly so that no unforeseen difficulties arise. Quite often, major equipment in cooling systems is only inspected at infrequent intervals. In some industries, the equipment is scheduled for turnaround or inspections possibly once every 2 to 3 years. Therefore, it is desirable to have some form or forms of monitoring results on an on-going basis. A few standard sidestream monitoring devices and test procedures will be discussed in this respect.

The most common method of monitoring overall results of cooling systems is to expose preweighed metal coupons in a bypass. Usually, four or more small metal coupons are exposed in a cooling water flow for various intervals of time. When the metal coupons are removed from the bypass, they are observed for possible slime accretions, deposition, and corrosion. If any pitting is present, the depth of the penetration can be measured with a microscope. Also, the coupon can be reweighed and the general corrosion rate calculated. Coupons of various metallurgy can be exposed to the cooling water flow.

Small test heat exchangers can also be installed in a bypass and used to reflect actual conditions on heat-transfer surfaces. With two or more test tubes in a small heat exchanger, the overall results in the system can be monitored without removing any major equipment from service. Heat-transfer coefficients and fouling factors can also be determined from test heat exchangers.

Some electrically heated single-tube test exchanges are available. Units of this type are usually equipped with a "see-through" outer plastic tube so that the test heat-exchange surface can be observed without difficulty.

Corrosion testing can be performed with the use of electrical instruments and probes exposed in the cooling system. Two types of electrical instruments are used, and these are based on electrical resistance and linear polarization resistance. The electrical resistance probe essentially gives the cumulative effect of corrosion, while the polarization techninque provides an instantaneous rate of corrosion.

For monitoring microbiological control programmes, exposed surfaces in the system, such as sumps and tower wood, can be observed for the presence of slime growths. The open decks of cooling towers can be examined for possible algae development.

Bacterial and fungal populations are counted periodically in some systems by standard serial dilution culture tests. Proprietary devices are also available to facilitate the counting technique. In some systems, the total aerobic bacterial population is counted by use of an appropriate culture medium. Biocides are then added to maintain the bacterial population below a certain figure, such as 100,000 per millilitre of circulating water. The threshold level for troublesome bacterial slime growth varies from system to system.

Chemical analyses of important operating parameters are conducted on a regular basis for overall control of a treatment programme. The. types of tests and the frequency of testing depend on the particular system. Quite often, testing three to four parameters every 8 to 24 hr is sufficient for control.

Deposits in the system can be analyzed for identification of chemical and biological components. Adjustments to the treatment programme can be made on the basis of deposit analysis. Microscopic examination of deposits can be used to determine the extent and types of organisms present. Cooling tower wood samples can also be examined in the laboratory to determine whether any chemical or biological degradation has taken place.

BIBLIOGRAPHY

Betz laboratories, Inc. (1980). "Betz Handbook of Industrial Water Conditioning," 8th ed. Betz Laboratories, Inc., Trevose, Pa.

Faust, S.D., and Aly, O.M. (1983). "Chemistry of Water Treatment." Butterworth, Woburn, Mass.

James, G.V. (1971). "Water Treatment," 4th ed. CRC Press, Cleveland, Ohio.

Hales, W.W. (1982). Use and reuse of water. Chemtech 12:532-537.

Pope, D.H., Duquette, D., Wayner, P.C., Jr. and Johannes, A.H. (1984). "Microbiologically Influenced Corrosion: A State of the Art Review." Materials Technology Institute, Columbus, Ohio.

"Steam: Its Generation and Use." 39th ed. (1978). Babcock and Wilcox Co., New York.

Strauss, S.D., and Puckorius, P.R. (1984). "Special Report: Cooling Water Treatment for Control of Scaling, Fouling, Corrosion," Power 128(6), S1—S24.

Vogely, W.A., and Mack, H.B. (1973). A 30 year review of water supply and demand. In "Complete Water Reuse. Industry's Opportunity" (Cecil, L.K., Ed.), pp. 12-27. American Institute of Chemical Engineers, New York.

6

WASTEWATER TREATMENT

Wastewater treatment is the means by which water that has been used and/or contaminated by humans or nature is restored to a desirable quality. Treatment may consist of chemical, biological, or physical processes or a combination thereof. Water may be treated to any level of quality desired; however, as its purity increases, so does the cost of attaining that purity. The required quality of water is dictated by its intended use, for example, aquatic life, drinking water, or irrigation.

I. WATER QUALITY MANAGEMENT

To understand the need for a particular degree of wastewater treatment, the concept of water quality management must be introduced. Water must be used and treated in such a way that deleterious effects are minimized, both for the environment and for the next user. From an engineering perspective, the intended use of a water should be determined, water quality requirements for that use should be delineated, and then treatment based on those requirements should be accomplished. [See *Environmental Engineering*.]

The dictionary defines pollution as something that makes the water physically impure, foul or filthy, dirty, stained, tainted, or demed. Actually, the addition of anything to a water that changes its existing quality constitutes pollution, including heat and sediment. Pollution may be caused by both humans and nature, and the effects may be identical. [See *Pollution Environmental.*]

The philosophy of water pollution control has changed considerably since the passage of the Public Health Service Act of 1912, which was primarily for the investigation of the spread of waterborne disease in people. Today, water quality standards are usually limited by aquatic life considerations. For example, the primary drinking water standard for zinc is 5 mg/litre, while concentrations of zinc as low as 0.001 mg/litre may be toxic to aquatic life. The implications of this difference are obvious as the cost to protect aquatic life is much greater than that for satisfying water quality standards for drinking water. It should be noted that the cost of removal of a contaminant increases considerably when the percentage of removal approaches 100 per cent. From the technical standpoint, water can be made as pure as desired: however, from the practical perspective, the cost of achieving a given purity must be considered. [See *Water Pollution.*]

A receiving water (lake, river, ocean, or estuary) has an innate ability to accept some contaminants without adverse environmental impact. This so-called self-purification capacity is denoted as the wast assimilative capacity. It may-defined as the amount of contaminant that may be discharged into a receiving water, under defined low flow conditions, that will not result in deleterious effects. This concept may very well dictate the wastewater treatment requirements imposed on a wastewater discharger. However, regulations or standards may dictate more removal than required to protect the environment. These are called technology-based standards and are based on what can be achieved technologically rather than what is needed environmentally.

II. WATER QUALITY

Before discussing the treatment of wastewater, it is instructive to establish the natural quality of water, that is, the natural constituents that would be expected to be found in water that has not been exposed to cultural activities or unusual environments.

The major sources of water are from rivers, groundwater, lakes, and impoundments. Inasmuch as water is known as the universal solvent, surface runoff may contain whatever the water contacts while seeking its sink. Thus, surface waters may contain high turbidity, taste, odour, and colour in addition to the many other contaminants that come in contact with the water. In addition, surface waters may be highly variable in their composition with respect to time.

Groundwater, however, can be expected to be free from organic odour and relatively constant in composition, and a good source will be constant in quantity. However, groundwaters may contain dissolved gases such as carbon dioxide and hydrogen sulphide, may be "hard" (*i.e.*, contain excessive calcium and magnesium), and may contain troublesome iron and manganese.

Thus, a typical water might contain relatively small amounts of calcium, magnesium, sodium, manganese, iron, and potassium as cations, and the anions might be sulphate, chloride, bicarbonate, and nitrate. Other materials found in the water could be considered as contaminants.

III. CLASSIFICATION OF POLLUTANTS

It is convenient to classify pollutants into four categories as follows: chemical, physical, physiological, and biological. A brief discussion of these is in order as is their method of removal. It should be noted that, depending on the intended use of the water, every receiving water will have a limit as to how much of each of these kinds of wastes can be discharged into the water without adverse effects.

Chemical pollutants can be broadly categorized into inorganic and organic pollutants, where organic materials may be defined as those compounds containing a carbon atom. The major problem with organic myterials is their conversion to carbon dioxide and water as follows:

$$\text{organics} + \text{microorganisms} + \text{oxygen} \xrightarrow{\text{nutrients}} CO_2 + H_2O + \text{microorganisms.}$$

Inasmuch as aquatic life requires a certain level of dissolved oxygen to live and propagate, it is obvious that if sufficient organic material is placed into the water, oxygen levels could be reduced to inimical concentrations. It is interesting to note that the process described is the same as that taking place in aerobic (presence of free oxygen) wastewater treatment processes. The common measure of oxygen depleting substances is the biochemical oxygen demand (BOD).

A prime consideration of inorganic chemicals is their toxicity. For example, changes in pH may occur from the discharge of soluble salts, and toxicity may occur directly from heavy metals. It should be noted, however, that organics may cause toxicity (*e.g.*, pesticides), and inorganics may cause oxygen depletion (*e.g.*, sulfurous acid). Also, all materials containing carbon may not cause an oxygen demand (*e.g.*, bicarbonate). [See *Water Pollution and Toxicology.*]

Other organic materials that should be mentioned include compounds of a toxic nature, such as pesticides, and taste-and odour-producing compounds, such as phenols and oils with their tendency to form surface films. The refractory organics are of concern because of the potential long-term cumulative effects of these materials in drinking water. While no adverse physiological effects have been reported, their safety is not assured.

Physical pollutants include colour, turbidity, temperature, suspended solids, foam, and radio-activity. While colour is not necessarily harmful, it may be aesthetically unacceptable for drinking water and some industrial uses. Colour is often caused by organic colloids, thus making its remaval expensive. Turbidity is caused by colloidal material and/or suspended solids, and its removal, like colour, may be expensive, requiring coagulation. Temperature is an important factor in biological activity and significantly affects chemical biological, and physical reactions. It may also act synergistically with toxic materials, for example, heavy metal toxicity increases with increasing temperature. Suspended solids, which may exhibit turbidity, can result from wastewater discharges or from natural processes, such as erosion. They may inhibit photosynthesis by reducing light penetration, decrease benthic organism activity by covering the water bottom with sediment, and interfere with fish activity by clogging their gills. Solids may be organic or inorganic. Foam resulting from surface active agents may cause esthetic problems, but developments by the detergent industry have minimized these effects. Surface active agents also may cause a reduction in the rate of oxygen gas transfer into the water. Radioactivity may be the result of fallout, natural sources, or waste discharges and can be incorporated into sludges or biological life or dissolved in the water. Because of the unique effects of radioactive substances, they must be controlled at the source.

Physiological effects of pollution are primarily the result of taste and odour. While taste and odour problems may be minor in effect, public reaction many times results in magnification of problems and concomitant adverse publicity of the water purveyor. Taste and odour are particularly objectionable when present in drinking water or process water for food where palatability is important. It is significant to note that phenol is detectable at concentrations of 0.001 mg/litre and is pervasive in the wastewater discharges of the petrochemical industry. The ability of taste-and odour-producing materials to taint fish is also important.

The biological category can also be divided into two subdivisions, public health considerations from waterborne diseases and eutrophication and/or biological growths resulting from nutrient additions.

The causative agents for waterborne disease include virus, protozoa, bacteria, and the helminths, and contemporary epidemic problems are from hepatitis (viralborne) and giardiasis (protozoan origin). Probably the most serious waterborne disease is cholera (bacterial origin) which caused major epidemics with high fatalities and morbidities in the 1800s in the United States.

The second category, which may be called secondary biological pollution, is the deterioration of water quality resulting from the addition of phosphorus and/or nitrogen to receiving waters which may be from wastewater discharges or from pollution with no single point source. When excessive biological growths and associated water quality problems occur in lakes and estuaries, the phenomenon is known as eutrophication. Eutrophication is a natural geological process that apparently may be accelerated by human cultural activities.

It should be noted that nutrient removal (*i.e.*, phosphorus and nitrogen removal) has been a major impetus for advanced wastewater treatment processes for discharges being made into affected lakes.

IV. CLASSIFICATION OF WASTEWATER TREATMENT

Wastewater treatment processes are categorized as pretreatment, primary treatment, secondary treatment and tertiary or advanced wastewater treatment.

Pretreatment is used to screen large solids, reduce the size of solids, remove floating oil, and equalize fluctuations in concentrations and/or flow. Pretreatment should include flow measurement.

Primary treatment includes pretreatment and refers to the removal of suspended solids, either by sedimentation or flotation. Chemicals may be added to aid particle removal and to neutralize highly acidic or basic influents. Primary treatment alone was common prior to the environmental movement, however, it is now unacceptable in the United States (1986). The only exceptions are for existing treatment plants along the coast that have extremely long discharge pipes, and they must demonstrate that the effluent causes no harm to the marine environment. It should be noted that the effluent from a primary plant, when discharging into a receiving water used for recreational purposes, is usually treated with chlorine for disinfection purposes prior to discharge.

Depending on the amount of organics contained in the solid material, primary treatment may remove a significant portion of the oxygen demanding substances (BOD). A well-designed and operated primary plant may remove as much as 35 to 40 per cent of the BOD and as much as 60 to 65 per cent of the settleable solids.

Secondary treatment adds a biological process after primary treatment which is commonly either activated sludge or trickling filtration or a modification thereof. These biochemical processes are aerobic and are the same as previously described as occurring in a river where organics are oxidized to carbon dioxide and water.

A well-operated and designed secondary treatment plant can be expected to remove 85 to 95 per cent of both BOD and settleable solids. Under existing regulations in the United States, all discharges must be subjected to at least secondary treatment except for those previously mentioned plants discharging into marine waters.

Tertiary treatment may be defined as treatment in addition to primary and secondary. It may include precipitation, filtration, coagulation and flocculation, air stripping, iron exchange, adsorption, membrane processes. nitrification, and denitrification. Those processes may be integrated into the secondary treatment plant or added onto the secondary effluent. Tertiary or advanced wastewater treatment can virtually attain any removal efficiency desired. However, as the percentage of removal increases, so does the cost of attaining it.

Special mention should be made of the solid material removed in any of these processes because solids or sludge disposal is a major problem in wastewater treatment. Sludges are particularly troublesome because of their high water content and their concomitant large volumes. Obviously, many of the processes mentioned in the preceeding paragraphs produce solids. A primary plant, for example, must have a means of disposing of the settled sludge and therefore may include an anaerobic sludge digester and sludge drying beds. Methods of sludge treatment include anaerobic (that is, in the absence of free oxygen) fermentation, anaerobic digestion, vacuum filtration, centrifugation, and filter pressing. The ultimate disposal is the real problem, however, because the choices are practically limited to land disposal, ocean disposal, or incineration. Each of these alternatives has problems which will be discussed later.

The processes mentioned are shown in diagrammatic form It should be noted that nitrification, denitrification, and membrane processes, which are not included, would fall into the category of tertiary treatment.

V. WASTEWATER CHARACTERIZATION

It is mandatory that the composition and characteristics of a wastewater be known and understood in order to design and operate a wastewater treatment plant. For example, the BOD is a measure of the organic material present and its removal depends on its form, that is, dissolved, solid, molecular type, etc. Solids characteristics will determine the sludge handling and transport facilities, pH may indicate the need for neutralization and oxygen content may demonstrate the need for odour control if a lack of dissolved oxygen is present. Grease and oil may cause problems and may require special removal facilities.

Wastes originating from industry may be toxic to biological processes, and thus their discharge to a municipal treatment plant is regulated. These may include heavy metals, hair, various chemicals, and grease and oil.

A major consideration in characterizing wastewaters is the flow since it may vary considerably and contain different materials at different times. For example, one would expect to find low flows occurring at night and peaks occurring during periods of maximum water use. Thus, the sampling of wastewaters should be done in accordance with flow so that the mass of constituents can be determined.

VI. OBJECTIVES OF WASTEWATER TREATMENT

The processes introduced will now be examined for their role in treating the various components constituting a wastewater. The prime objectives are twofold:

1. to separate the solid from the liquid fraction and concentrate the solids collected from the carrier water, and

2. to remove or render inoccuous materials that will cause adverse effects when the effluent is subjected to ultimate disposal.

It is instructive to note that only 0.1 per cent of domestic sewage is solids, the remainder being the carrier water.

In general, less than 50 per cent of the waste material in domestic sewage remains in suspension, thus allowing separation by straining, skimming, or settling. The remainder must be destroyed or rendered removable. This is accomplished by biological, physical, or chemical means.

In chemical treatment, the coagulating chemicals combine with the finely divided and colloidal (nonsettleable) material to form settleable flocs. In aerobic biological treatment, living organisms consume finely divided colloidal and dissolved decomposable substances and convert them into carbon dioxide and water and settleable films, slimes, or flocs, primarily consisting of cell material.

Physical treatment may involve absorption processes, where, for example, material containing a high affinity for activated carbon may be removed.

VII. PRETREATMENT

A. Screening

Large, objectionable, solid material must be removed from a wastewater prior to its entry into the main part of the treatment plant. This is accomplished by some type of screening device which usually consists of bar screens. Since the solid material not passing through the screen will be retained, the flow will eventually be hindered. Thus, provision for removal must be made and is accomplished manually in small plants and continuously in larger plants.

Note that a by-pass is provided in case the flow is accidentally obstructed to the point where the flow cannot pass through the screening device. Screenings are putrescible and objectionable and are usually buried or incinerated.

B. Comminutor

A comminutor or large "garbage grinder," may be used to homogenize the solids so that subsequent treatment is more effective. These devices are generally placed between the grit chambers and the primary settling tanks to protect the machinery from the gritty material.

C. Oil and Grease Removal

Excessive oil, grease, and finely divided suspended solids should be removed prior to discharge into the primary sedimentation tank. This can be accomplished by processes including skimming, flotation, and aeration. When air bubbles are introduced into a tank, the solid particles tend to adhere to the bubble and float to the surface where they can be removed by skimming. Flotation can also be accomplished by introducing air under pressure into the water and then allowing the pressurized air to rise to the surface or by saturating the water with air, applying a vacuum, and allowing the air to escape with the removable material attached.

A schematic diagram of a dissolved air flotation unit is demonstrates the method by which the removable particles float to the surface and are removed from relatively quiescent water zones.

D. Grit Removal

Grit is considered to be composed of small coarse particles of sand, gravel, or other minute mineral material. Grit removal should be introduced prior to any mechanical equipment to minimize abrasion. Grit removal facilities usually consist of either an aerated chamber, where the air supply is just sufficient to keep the organic material in suspension and allow the heavier material to settle. Grit is also removed by adjusting the velocity of flow through a chamber so that the gritty material will settle and the organics will remain suspended. Unless the grit is washed, it will contain organic material and must be buried at a solid waste disposal site.

Note that an advantage of the aerated device is maintaining freshness of the sewage because of the air introduced, and an advantage of the velocity control chamber is the ability to measure flow because of the head loss induced by the velocity control device.

E. Equalization and Neutralization

It is sometimes necessary to install a basin to neutralize large fluctuations in pH and/or concentrations of contaminants in the incoming flow to the treatment plant. Also, the previously mentioned fluctuations in volume with time may necessitate such a holding tank to equalize the flow. The use of this type of chamber will assist in maintaining a relatively uniform flow and concentration through the plant and to the receiving water.

VIII. PRIMARY TREATMENT

The settleable solids are removed by introducing the wastewater, after pretreatment, into a large rectangular or circular tank where use is made of gravity settling. The removal process may be illustrated by considering placing the wastewater in a container of the same depth as the sedimentation tank. If the container were moved from entrance to exist at the same velocity as the water flowing through the tank, the solids collected in the container at the end of the tank would be proportional to those collected in the sedimentation process itself. This assumes ideal conditions in the tank, however.

Since solids will be collected at the bottom of the settling tank, provision must be made for their removal. This is usually accomplished by a continuous belt device. The solids of sludge are then pumped from the sludge hopper to a sludge digester or other sludge treatment facility. The secondary tank would be used subsequently to biological treatment, and scum removal would probably not be required. It should be noted that chemicals are sometimes added prior to settling to enhance solids removal.

The combination of pretreatment and sedimentation is known as primary treatment, and prior to the 1970s, it was common to chlorinate the effluent and discharge it into the nearest waterbody. As previously mentioned, primary treatment alone is no longer acceptable except for a few long ocean outfalls which must obtain a special variance.

IX. SECONDARY TREATMENT

Secondary treatment utilizes some form of biological treatment subsequent to primary treatment. Since much of the organic material in a wastewater may be colloidal or dissolved, the processes described thus far will be ineffective in its treatment. It has previously been shown that organic matter will be oxidized to carbon dioxide and water in the presence of microorganisms, oxygen, and nutrients. Thus, the only conditions required for wastewater treatment are an adequate number of microorganisms that like to eat the organic material, an oxygen supply, nutrients, a means of intimate contact between the microorganisms and the food (wastewater), and a method of containment.

The process is identical to that occurring in an aerobic stream except that in the wastewater treatment plant, the process is controlled, condensed in time, and continuously supplied with oxygen to maintain an aerobic state.

There are numerous ways to design a biological wastewater treatment process; however, the two most commonly used are the activated sludge process and the trickling filtration process. In the activated sludge process, the fine, suspended, and colloidal and soluble organic materials are brought into intimate contact with a biologically active sludge maintained in suspension in the tank by introducing air that serves not only to maintain turbulence and maximum contact, but supplies the microorganisms with the oxygen they require for their metabolism. The activated sludge performs the work of adsorbing, assimilating, and flocculating the waste material.

The trickling filtration process is identical to the activated sludge process in principle. However, instead of the water flowing through the suspended sludge containing the micro-organisms, the waste material flows over a suitable surface to which the microorganisms adhere. When biological material flows over a suitable surface, growth of bacteria and other microorganisms will occur. These microorganisms create a slimy, gelatinous film that transfers the matter held in suspension, both colloidal and in solution, to the microorganisms, which takes out substances needed for their growth, and transfers back to the liquid the end products of decomposition, including (in the aerobic process) nitrates, carbon dioxide, and sulphates. As is the case with the activated sludge process, the microorganisms working in the trickling filtration process need only a continuous food supply, adequate oxygen, a suitable supporting film or floc, and appropriate nutrients. The processes taking place in any aerobic biological treatment plant can best be described schematically as shown in Figure 6.1, where both cell material and end-products are produced by the bacteria, depending on the existing ratio of food to microorganisms in the process. Thus, part of the energy may be utilized in reproducing new bacteria and the rest of the energy used in converting waste products to, for example, carbon dioxide and water. If the food supply becomes meager, the bacteria will eat themselves and bacterial cells will be converted into end products. This process, when the food-to-microorganism ratio is low, is called endogenous respiration. It can also be seen from Figure 6.1 that part of the waste material will be nonbiodegradable. This is the residue of the waste material that will not be removed, either as cell material or in the end products, and is the residue of the process. The endogenous respiration process is analogus to a starving human when he lacks food, because he then uses his own tissue to provide energy to sustain his life processes.

Schematically demonstrates a biological wastewater treatment process, where Q is the flow volume, S is the concentration of organics, V is the volume of the tank, M is the mass of microorganisms in· the tank, s is the concentration of organics in the tank, and R is the fraction of return flow.

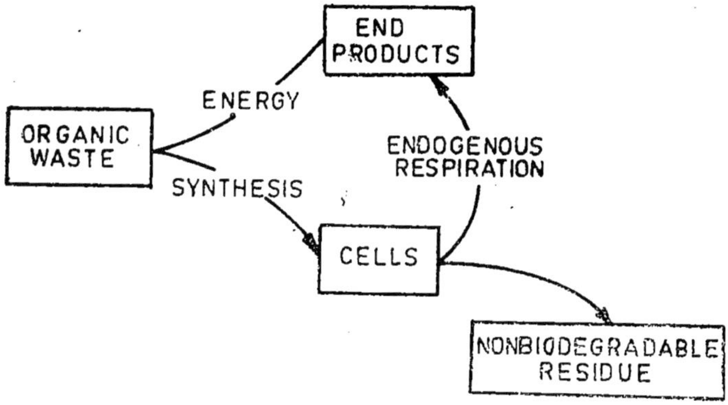

Fig. 6.1: Biological stabilization of organic waste materials

A. Activated Sludge Process

The activated sludge process consists of maintaining an active floc in a tank supplied with oxygen so that maximum contact is made between the incoming wastewater and the microorganisms in the floc. In the conventional process, a rectangular tank is usually used and the wastewater is introduced into a concentration of microorganisms maintained in the tank. Air is introduced. either in the form of bubbles through diffusers or by turbulent agitation of the liquid by an impeller. The process is maintained in the tank by returning a certain portion of the sludge that passes through the tank and is settled out in a subsequent or secondary sedimentation basin. Remember that the activated sludge process produces new cell material which will become part of the activated sludge mass. Part of the settled material must be disposed of, and a portion must be introduced into the incoming raw wastewater in order to have an active population of microorganisms that will feed on the organic compounds.

As might be expected, the design of an activated sludge process will depend on the ratio of the food, or waste, to the microorganisms, or activated sludge. Interestingly enough, most organic materials are adsorbed by the sludge floc in 15 to

45 min, although most conventional plants are designed with at least 30 to 90 min contact time for adequate adsorption by the sludge floc. With the activated sludge process, it is possible to obtain removals of BOD on the order of 95 per cent and a clearer effluent than with most other biological oxidation processes. Many modifications of the activated sludge process exist, including a high rate process where the food-to micro-organism ratio is high, thus producing more sludge; a step-aeration process, where the influent is added at intervals along the aeration tank; tapered aeration, where the introduction of air is varied along the length of the tank, the higher concentration being at the influent point of the tank; and the contact stabilization process, where a portion of the sludge is aerated separately, thus adding flexibility to the process.

A modification of the activated sludge process is called extended aeration, where it is assumed that the system is completely mixed and exceedingly long detention times are given to the aeration operation. This means that a high solids content or a low food-to-microorganism ratio will exist. Thus, endogenous respiration of the sludge will occur and the sludge will "burn itself up." The process is sometimes called complete oxidation, although there will always be some biological residue, inorganics and solids that will necessarily emanate from the system. The process of extended aeration is frequently used in small installations such as schools, subdivisions, and models and some industrial plants.

B. Trickling Filtration

As previously mentioned, the trickling filtration process is identical to the activated sludge process, except that the micro-organisms working to stabilize the organic waste material are attached to a fixed bed rather than being of a floating, suspended nature. The wastewater is distributed from rotary nozzles over the bed which usually consists of coarse, rough, hard material that gives support to the biological film. The organics in the wastewater are oxidized after assimilation by the bacteria and then released in a more stable form. Periodically, the film will become so thick that it can no longer be supported on the media and will slough off and be discharged into the effluent from the filter. This sludge will then go to the secondary settling tank and be removed with that from the primary settling tank. The trickling filtration process is said to be advantageous because it can provide good performance with a minimum of skilled operator attention. However, the process is highly temperature dependent and will not always perform in accord with the requirements of present-day regulatory agencies. Note that recirculation of the effluent into the influent is commonly practiced.

Oxygen is supplied because of the "open air" nature of the filter, and the wastewater is applied to the filter media via the distributor, which rotates because of the momentum induced by the carrier water. The media shown is stone, although many filters utilize synthetic media such as PVC, which allows greater depths of filtration.

C. Lagoons and Oxidation Ponds

Stabilization ponds are probably one of the oldest wastewater treatment processes in existence and are still used in many locations; about 7000 exist in the United States today. They can be used alone or in combination with other wastewater treatment processes, their design and utilization depending on many factors, including weather, land availability, purpose, and location. Their detention time varies from 7 to 180 days, depending on the type of pond.

Ponds can be classified into four categories: facultative, aerated, aerobic, and anaerobic, the most common being the facultative pond.

The facultative ponds are usually 4 to 8 ft in depth and operate aerobically in the upper layers and anaerobically in the lower depths. Oxygen in the upper portion is supplied by photosynthesis and surface reaeration. The algae present in the effluent may create problems, however. The detention time of a facultative pond may be from 5 to 30 days. Effluent BOD values range from 20 to 60 mg/litre and effluent suspended solids from 30 to 150 mg/litre.

Aerated ponds are supplied with oxygen by mechanical or diffused aerators and are 6 to 20 ft in depth with detention times of 3 to 10 days. Their performance can approach that of activated sludge; however, energy costs become high and secondary sedimentation with sludge return must be utilized.

Aerobic ponds are about 12 to 18 in. deep and maintain oxygen throughout their depth. The oxygen is supplied by photosynthesis and surface reaeration which is sometimes aided by mixing. These high rate ponds are limited to warm, sunny climates and have a detention time of 3 to 5 days. In addition, unless algae removal is practiced, the effluent will contain a high percentage of suspended solids.

Anaerobic ponds have such heavy loading that no aerobic zone exists. They are usually used for strong industrial or agricultural wastes or as pretreatment. They are 8 to 15 ft deep and have detention times of 20 to 50 days. Obviously, a major problem is with odour produced from the anaerobic process.

D. Rotating Biological Contactors

The rotating biological contactor (RBC) operates on the same principles as activated sludge and trickling filtration and has several advantages that have resulted in an increasing number of installations in the United States.

The process consists of large diameter plastic media mounted on a horizontal shaft that is slowly rotated in a tank containing wastewater. The biological slime develops on the media, which has approximately 40 per cent of the surface area submerged. As the discs rotate, the wastewater trickles down the face of the contactors, providing the necessary reaction time for aeration and metabolism of the attached microorganisms. Shearing forces exerted on the biomass cause sloughing from the discs, which is removed in a secondary clarifier.

With an RBC, BOD removal and nitrification may be accomplished in the same tank. In addition, recycling is not necessary and the power costs are considerably less than for activated sludge. Denitrification can also be accomplished using RBCs.

E. Septic Tanks

A discussion of biological treatment would not be complete without mentioning the septic tank, inasmuch as a large segment of the population is still served by this primitive device. From previous discussion, it can be seen that the process is anaerobic, will accumulate solids, and unless periodically pumped out (every 2 to 3 yr), will discharge objectionable material. The discharge from such a tank is usually into a designed drain field and ultimately into the groundwater and/or a surface water. It is obvious that a well-designed drainage system into a suitable soil is mandatory.

X. SOLIDS (SLUDGE) TREATMENT

One of the most pervasive problems in water pollution control is the disposal of the solids produced in the various separation processes. If chemicals are added, the sludge produced may be different from a biological sludge and will increase the volume of sludge produced. Thus, the nature of the solids produced may vary considerably and therefore necessitate different treatment prior to its ultimate disposal.

The choices for the ultimate disposal of sludge are few and consist of some kind of land application, incineration, and ocean dumping. While ocean dumping is still practiced by some large cities, it is now prohibited and no longer a viable alternative. Also, incineration leaves a residue that must be disposed of. Thus, the ultimate residing place of the solids produced in wastewater treatment is into or onto the land.

The objectives of the processes used to treat sludge prior to its final disposal are to reduce the volume, destroy pathogens, remove water, improve efficiency of subsequent processes, control putrescibility, and stabilize organics. Because the final product must be transported to its ultimate disposal site, attainment of these objectives will reduce transportation costs and minimize adverse environmental effects.

XI. ULTIMATE DISPOSITION OF SOLIDS

As previously mentioned, because of adverse environmental impacts, the ultimate disposal of solids or sludge must now be onto or into the ground. The options available are to use the sludge for agricultural purposes, to apply sludge to forested land, to use it for land reclamation and is presently quite common in swimming pool filtration.

Chemical clarification consists of four phases; coagulation, flocculation, sedimentation, and filtration. Coagulation, essentially the addition of chemicals that are rapidly mixed with the water, usually involves the use of the oxides of aluminium, iron, or calcium. Once the coagulation step is accomplished, the fiocculation process then allows aggregation, neutralization, and adsorption of the floc particles. This process then leads to the sedimentation process, where the flocs that have been previously formed are allowered to settle to the bottom of a settling tank. While most of the flocculated material will be removed in the sedimentation tank, it is usually required to further remove the floc particles that do not settle by the use of a filtration process which is conducted in beds of porous media such as sand.

The usual means of filtration has been through sand beds containing graded sand placed on a supporting medium and containing an under-drain to collect the filtered effluent. It is obvious that as wastewater containing solids is passed through this type of filter, the solids will accumulate and eventually clog up the openings. Thus, some provision must be made to remove the collected material. The procedure usually followed is to backwash the sand, that is, to reverse the flow so that the sand is suspended and the lighter material is washed away. It should be noted that subsequent to a backwashing operation, the sand will settle according to size and thus become stratified, with the smaller particles at the top. Since these particles will filter out most of the solid material, it may be concluded that the entire depth of a sand bed is not utilized in a filtration process. To alleviate this problem, a multimedia filter has been developed, involving different filtration materials, each having a different specific gravity. Materials commonly used are sand, coal, and garnet. This type of filter extends the length of the filtration runs and demonstrate a high efficiency.

The moving bed filter is a new technique that is essentially a form of countercurrent extraction, that is, feeding the sand countercurrent to the filtering water. As the filter surface becomes clogged, the filter medium is moved forward and a new surface is exposed. It is said that, theoretically, 1 per cent of the filter is being backwashed 100 per cent of the time as compared to a conventional backwashing operation where 100 per cent of the filter is backwashed 1 per cent of the time.

Probably the newest filtration concept to be investigated is ultrafiltration, which is similar to reverse osmosis, but differs with respect to the size of the particles that are separated. Ultrafiltration will generally not retain a particle whose weight is 500 to 1000 times the molecular weight, and thus it fails to separate the inorganic salts. The pressures used in ultrafiltration are on the order of 50 lb/in.^{2} as contrasted to pressures exceeding 500 lb/in.^{2} in reverse osmosis. The process is essentially one of pressure filtration, using semipermeable membranes that act as molecular screens and separate colloidal and molecular materials that are dissolved or suspended in the liquid phase.

B. Organic Removal

A previously indicated, there are measurable amounts of refractory soluble organics, or those materials that are not removed by inventional processes, remaining in a secondary effluent. The composition of these materials is not always precisely known. Although they are not easily biologically oxidizable, they may exercise a BOD in a receiving water. In addition, these materials may cause taste and odour problems in a water supply, and more important, the potential toxicological effects on humans have not been precisely delineated.

The process of removing these materials uses activated carban, which possesses the property of attracting many contaminants to adhere to its surface. The phenomenon is called adsorption. Activated carbon can be made very porous, increasing its surface area to unit weight ratio enormously and can tous have a high adsorption capacity for the soluble organics in wastewater. It can be put in contact with the wastewater either as a very fine powder or as granules. Since the powdered form is not conducive to column operation, granular activated carban is used most frequently in advanced wastewater treatment plants. Obviously, as the wastewater passes through the bed and the constituents from the wastewater are absorbed onto the carbon, the sites available for adsorption will be exhausted and the bed must be replaced. The replacement can either be by new activated carbon material or by thermal regeneration of the exhausted bed. In large plants, regeneration by high températures is usually practiced.

C. Inorganic Removal

Many of the inorganic constituents of a water will be at a higher than desirable concentration in a secondary treatment plant effluent. While in most cases these excessive concentrations will not be harmful, continuous reuse of the water will cause them to build up rapidily to prohibitive levels. Thus, a means of removing inorganics via a tertiary treatment process may be necessary.

Ion exchange is a process that has been used for many years in the removal of hardness from water and of specific ions such as certain metals. The process takes advantage of the ability of certain natural and synthetic materials to exchange an ion contained in the material for an other contained in the water passing through it. For example, if the passed a water containing calcium ions through an ion exchange media that preferred calcium ions instead of the sedium ions that were already attached to it, the calcium ions would adhere to the exchange media and the sodium ions would pass out into the effluent. As with adsorption on carban, if a solution containing an ion to be removed is continuously applied the the ion exchange medium, the medium will eventually become exhausted. It may then be subjected to a process called regeneration, where the exchange medium is converted back to its original state by passing a chemical through it. There are many modifications and combinations of ion exchange units that have been used in both water and wastewater treatment technology. It should be noted that it is possible to recover material from an ion exchange unit that might be economically advantageous. The use of ion exchange media probably most familiar to the layman is the device that can be placed on the faucet in the home to obtain soft water for steam ions and other uses.

The electrodialysis process is used for demineralizing wastewaters. When voltage is applied across a cell containing mineralized water, the negatively charged ions (anions) will migrate to the positive electrode and the positively charged ions (cations) will migrate to the negative electrode. Aalternate placement of anion and cation permeable membranes causes alternate compartments to become more concentrated in salts while others become more dilute. It should be noted that both suspended solids and dissolved organics should be eliminated from the feed water to the electrodialysis unit because of the possibilities of plugging and fouling.

Theoretically, the process of reverse osmosis is capable of removing greater than 90 per cent of certain inorganic ions, organic material, and colloids. The Principle involved is demonstrates various osmotic systems leading to the reverse osmosis process, which shows the concentration of a spiral-wound reverse osmotic module. In operation, the water containing the dissolved materials is placed in contact with the membrane at a pressure in excess of the osmotic pressure of the solution. Under these conditions, the water will permeate the membrane and concentrate the dissolved materials. Problems with the reverse osmosis process have primarily been with the membranes because they are subject to fouling.

D. Nitrozen Removal

As previously mentioned, the presence of the various forms of nitrozen in water can lead to several problems, including entrophication, BOD in a receiving water, and methemoglobinemia in infants. The removal of nitrozen may need to become an integral part of a tertiary treatment plant.

Successful removal of ammonia nitrozen has been accomplished by raising the pH level about 10 and removing the converted ammonia by aeration. The pH is usually raised by the use of lime and a packed tower is used for the gas stripping operation.

Nitrozen can also be removed by selective ion exchange, where ammonium ion is removed by using clinoptilolite, a naturally occurring medium.

The biological process used for nitrozen removal is attained by modifying the activated sludge process to increase the oxidation of organic nitrogen to nitrate and then subsequently denitrifying the nitrate by placing it under anaerobic conditions and allowing the bacteria to conver the nitrate to nitrozen gas. Both of these processes have been used successfully. However, the bacteria involved are highly sensitive to temperature and this limitation may preclude practical biological nitrification in cold wheather. The conversion of ammonia nitrogen to nitrate is called nitrification, and nitrate to nitrogen gas conversion is called denitrification. The same reactions occur in surface waters and groundwaters.

It is also possible to oxidize the ammonia nitrogen with chlorine, a process called breakpoint chlorination. While expensive, it has the advantage of eliminating the ammonia and not producing nitrate, thus precluding the need for denitrification. The same advantage exists for air stripping.

E Phosphorus Removal

Because of its ability to stimulate biological growth, phosphorus in wastewater effluents is limited to very low levels whenever the receiving water discharges into or is a lake or impoundment. The major source of phosphorus in domestic wastewater is detergents, and some states in the United States and provinces in Canada have controlled phosphates in detergents.

Phosphorus removal can be accomplished by chemical precipitation and by a modification of biological oxidation referred to as "luxury uptake." Activated sludge has also been